Understanding Evolution and Ourselves

reviews by

Dennis Littrell

Dennis Littrell

This book is dedicated to the memory of Charles Darwin, one of the greatest scientists to have ever lived.

Dennis Littrell

Table of Contents

Dennis Littrell

Introduction

I've always had a keen interest in evolution, and beginning in 1999 I started to review the books on evolution as I read them, and then posted the reviews at Amazon.com and elsewhere on the Web. Part of my motivation for writing the reviews was to improve my understanding of what I read and to keep a record of my readings. I was encouraged by readers who found that my reviews were often insightful and educational in and of themselves. One reader went so far as to say that my reviews so well captured the essence of the books that one could just read my reviews and skip reading the books! I would advise against that of course, although there is something to be said for really good "Cliff Notes."

The value of this book, and the primary reason for publishing it, is not so much in pointing to and evaluating various good and not so good books on evolution, but in the pure didactic quality of the reviews. Simply put I believe that the reader can gain a very fine introduction to the very complex subject of biological and cultural evolution by reading this book.

Most of the books are by professionals in one or more of the biological sciences or by science writers who are expert in those fields. Some of the books are written by psychologists, paleoanthropologists and others from allied disciplines. A few are written by non-scientists who provide a view of evolution from outside academia.

My position as a reviewer has been to be critical when I think criticism is due and to be appreciative when I think it is warranted. I usually write positive reviews since any book I read all the way through can't be without merit, although there are exceptions. I will leaf through several books without reading them after looking at many titles. I have read so many books that it doesn't take long to know whether a particular one will be worth reading, especially in a subject like evolution with which I am very familiar. It should be noted that I am in a somewhat enviable position since I don't have to maintain a position within an academic or professional community. This leaves me free to say what I think candidly.

The reviews are at most one thousand words long (since that was the word limit imposed by Amazon.com). They are presented in alphabetical order within each chapter. The number of stars following the titles, one to five stars, five stars being best, indicates how much I liked the book and/or how valuable I thought the book was.

I have corrected some typos and some errors of fact, and where I have changed my mind or learned otherwise, I have made a note. In most cases where perhaps today I would phrase something differently or make a different emphasis I have left the review as I wrote it.

Chapter One

Human Evolution

The concentration in this chapter is on books that shine light on how we are different from others hominids and how we became anatomically and behaviorally human. We know that our ancestors branched away from the ancestors of chimpanzees (our closest living DNA relatives) something like six million years ago, but the precise linage of proto-humans and/or hominids that led to us is unclear and likely to stay that way for some time to come.

We migrated out of Africa, but exactly when or which direction we took first is unclear. Did we make it to the Mediterranean and then turn right or did we cross the Red Sea near the Gulf of Aden during a glacial period when the sea level was low, going east into what is now Yemen?

We know that behaviorally modern humans replaced the Neanderthal in Europe, but just why or how is unclear. Did we out-compete them in the cold because we had needles and such to sew skins tightly around our bodies, and other cultural advantages, or did we kill them with our superior weaponry or aggressiveness? Or did they die from diseases we brought with us as did the natives of the Americas tens of thousands of years later? Or (most likely) was their demise the result of a combination of factors?

What did our ancestors eat and what difference did that make? Why did we evolve to walk upright and what are the advantages of walking upright? Why and how did our big and calorically expensive big brains evolve?

These and other questions are considered in the books reviewed in this chapter.

Adovasio, J.M., Olga Soffer and **Jake Page** *The Invisible Sex: Uncovering the True Roles of Women in Prehistory* (2007) ****
Very interesting and readable with some avoidable sexism

J.M. Adovasio is an archaeologist. Olga Soffer is an anthropologist, and Jake

Page is a science writer. They have put together in "The Invisible Sex" a book that attempts to

(1) Bring the general reader up to date on the latest developments in archaeology or paleoanthropology;
(2) Uncover the True Roles of Women in Prehistory (as in the subtitle); and
(3) Provide a corrective to a male-dominated view of the prehistory.

The main image they want to correct is that of the great male hunter bravely slaying mastodons and in general bringing home the bacon to an adoring and appreciative family or band. What the authors want readers to see is that women weren't just tag-alongs on the way to our becoming fully modern humans, but at least equal partners. The authors refer to nets, threads, garments, basket weaving, cordage, digging sticks, the famous "Venus" statuettes, and other cultural artifacts to demonstrate the enormous role that women played culturally. They speculate that women invented farming, that they too engaged in the hunt, as well as producing works of art as important as the famous cave paintings.

The main method used by the authors is to infer the past from a study of recent hunter-gatherer societies while comparing ancient artifacts with more recent ones. This method certainly ought to provide insight into human life in prehistory, but of course there are some problems. The main one I think is that the "primitive" societies extant today or in the near past are not necessary typical of those that existed in prehistory because today's tribes occupy marginal lands since the best lands have long been given over to modern societies.

Personally, I never had any doubt about the significant role females played in the history of the species. Indeed, my feeling has always been that women are the default human being, and men an appendage, a necessary evil if you will. (Ha!) I don't think we need to study archaeology to understand that the central role in human culture is and was occupied by women. There is a sense of pandering and begging the question in the way the authors insist on the obvious. I think it stems from the fact that women in some of the sciences have felt and still do feel like second class citizens.

But that is changing. As the authors point out, most anthropologists today are women. The old male-delusional interpretations of culture in paleo-societies or in modern gatherer-hunter societies are a thing of the past. Instead we are in danger of having female-delusional interpretations. Here are a couple of examples of "reverse" sexism in the text:

From page 209: The authors imagine that "Aboriginal men" may have sniffed "contemptuously at the shell hooks and...strings that their women were using, making invidious comparisons of those little toys...with their mighty, multi-pointed, barbed, aerodynamic spears and other large instruments." Actually the men may have looked admiringly at such tools since such tools increased their subsistence.

On pages 248-249 in pre-Columbian New Mexico: While the women were farming, "The men had continued to spend much of their time roaming the surround, hunting (or goofing off?)." I think time spent "goofing off" applies to both sexes.

Frankly I am a bit weary of books that focus on sexism in one form or the other to the exclusion of the science itself. This book would have been a lot better had the stance been devoid of sexism and just concentrated on what the authors have learned and understand. Their various interpretations of the enigmatic Venus of Willendorf figurine, from goddess to porn star, is a case in point. Clearly the figure, which the authors quite naturally attribute to a female artist, is a symbol in some sense of fertility, not just the fertility of the female, but of the earth itself since no woman could have gotten so corpulent except during a period of plenty. And that is what probably enamored those who made and kept such figures—the idea of the season of plenty. Such a woman not only had plenty to eat, but was a heavy favorite to survive whatever winter may come. Her personal sexuality is secondary to the generalized idea of fertility.

As for bringing the general reader up to date on the latest developments in archaeology or paleoanthropology, the authors provide some interesting material. What has happened is that because of new technologies and more professional care taken by the scientists themselves, we are now able to unearth and be aware of artifacts such as threads, baskets, nets, etc., in a way previously not possible. And, it is true, it helps to see these artifacts from a woman's point of view, that is, as a gender female looking at what happened and assessing the importance of the artifacts, and drawing conclusions that did not occur to the old guys who once dominated the social sciences. Of course even better would be a balanced perspective, a fully *human* perspective, but we still have a ways to go to achieve that.

Perhaps the most glaring omission in the book is the failure of the authors to mention war (or what I like to call "the war system") as a reason for the rise of patriarchy during the transition from mostly hunter-gatherer societies to agriculture ones. Before there were storehouses of grain and large settled communities, the profits of war were meager. Once war became a viable occupation, men increased their power over women. Indeed the current religions of the Middle East, Christianity, Judaism and Islam, are all warlike and patriarchal in their conservative forms.

So indeed, the authors do help uncover the true roles of women in the prehistory for those of us who had any doubt. However, whether women went on the Big Hunt or not, or whether men ever acted as "midwives" (which the authors identify as the real "oldest profession") is of secondary importance to the fact of hunting and midwifery.

Allport, Susan *The Primal Feast: Food, Sex, Foraging, and Love* (2000) *****
A multi-dimensional look at food and how it has shaped us

Allport sees herself as a forager, a creature with a drive to look for food. She attributes this drive to her ancestors who spent much of their time searching the forests and savannas for food. From this personal observation, keenly felt, Allport branches out to thoughts about food and eating, from the habits of the deer and squirrels near her home to the proclivities of the chimpanzees of Africa. Primary among her concerns is how these behaviors relate to human food consumption, and how the search for food and what we eat has shaped our social structure and psychology.

This is a very interesting read, graceful written and full of intriguing bits of information. Did you know, for example, that virtually all common spices, oregano, thyme, cinnamon, rosemary, etc. have "powerful antibacterial and antifungal effects" (p. 118)? Or that there is a beeswax-eating bird used by the Hadza people of Tanzania that leads them to bee hives? The bird loves beeswax but is unable to open a hive, but is rewarded when people do. This "honey guide" is thought to be human kind's "oldest surviving partner in predation, much older than the dog or the falcon" (p. 148). Or that corn treated with an alkali (tortillas are made with lime) frees the otherwise unavailable essential amino acid tryptophan from the corn so that those who depend heavily upon corn in their diet will not develop pellagra, an often-fatal dietary disease? This is just one example of an eating technique developed through trial and error and happenstance that allows a people to live on an otherwise incomplete diet—a "cuisine" altered only at considerable risk.

Allport also goes from observation to speculate on such things as the origins of tool use, the sexual differentiation of hunting and gathering, and the use of food for social and sexual advantage. Generally she follows the well-documented and successful path of evolutionary biology and psychology, noting along the way where earlier ideas have proven wrong or incomplete (Raymond Dart's mistaken belief that Australopithecus was largely a meat-eater (p. 157) is a case in point.) She is insightful and presents her arguments well so that we tend to agree with what she says. Her idea that tool use began with females and then later spread to males, as presented in Chapter Twelve "The Nature of Food," is persuasive. Particularly interesting to me is the material on the nature of omnivores and how food choices dictate physiology and vice versa. For example, primates with their big brains that require large amounts of energy rich foods cannot subsist on leaves and other foods requiring long intestinal tracts and a slow-motion life style. Or, reverse that and observe that creatures that have the ability to find and consume energy rich foods can grow big, energy-demanding brains, while those who eat leaves and other foods that require a lot of digestion can't afford to grow a big brain.

Also interesting is the chapter on food and cooking aptly entitled, "The Only Cooks on the Planet." Cooking and other processing techniques such as leeching and preserving freed up many foods for our consumption not available to other creatures. In this connection, Allport makes the astute observation that the technique of cultivation, that of agriculturally engineering energy-rich and less toxic foods, made these plants edible to other animals creating a new ecology of vermin (p. 124). On the other hand the technique of cooking makes foods available only to humans. Well, again except for some very selective vermin.

One of the more startling observations made by Allport, who really has a keen eye for connections, is this on page 60. She is discussing the differentiation of sex cells, the female cell stationary and energy-rich, the male mobile and with little nutritive value. She quotes biologist Robert Trivers as saying, "An undifferentiated system of sex cells seems highly unstable." She concludes, "So as soon as selection favored those that invested their sex cells with nutritious substances, it also favored those that cheated the system and became adept at numbers and mobility instead. As soon as selection favored eggs, it also favored sperm. And there you have it: the origin of the sexes."

This is startling because biologists are still in a quandary about how sex began. The main and latest idea has been that sexuality developed as a result of the arms race between the organism and its microbial predators (see Margulis, Lynn and Dorion Sagan. *Mystery Dance: On the Evolution of Human Sexuality* (1991) or Matt Ridley's *The Red Queen* (1993) for examples of this argument). Here however Allport suggests that one of those predators may have been another cell bent not on consumption, per se, but on reproduction! And so they formed a symbiosis...

I am pleased to note that although Allport doesn't mince words when it comes to pointing out male maleficence—apt and hard-hitting is her discussion of how in many cultures males manufacture food taboos that limit the foods females can eat, saving the biggest and best portions for themselves—she plays fair throughout, and at no time gets bogged down in the sexism that preoccupies some writers. On page 190, for example, she states quite directly that females shape male behavior by their reproductive choices, thereby implying that females are also responsible for the male violence that we post-moderns so wisely abhor.

Allport appropriately ends the book with a plea that we not turn the planet into "a giant McDonald's dispensing Happy Meals" to "Homo Sapiens alone," and that we not overuse the world's resources. Amen to that, and kudos to Susan Allport for writing such an interesting and wisdom-filled book.

Arsuaga, Juan Luis *The Neanderthal's Necklace: In Search of the First Thinkers* (2002) *****
The Neanderthal as a nearly contemporary, parallel species

This is a fine book that sheds further light on what the Neanderthals were like and what happened to them. Written in an engaging, clear and almost poetic style (the translation by Andy Klatt is first rate, his surprising use of "irregard-less" on page 182 notwithstanding), this book gives us a sense of the latest understanding with an emphasis on evidence from the Sierra de Artapuerca excavation site in Spain where paleoanthropologist Arsuaga is co-director. The black, white and gray illustrations by Juan Carlos Sastre nicely augment the text.

Professor Arsuaga begins with the observation that today we exist almost alone in the sense that there are no very similar species extant, the last one being the Neanderthal in Europe. Arsuaga then traces our descent until he arrives at "Domesticated Man" in the Epilogue. His detours and asides are very interesting. I was especially pleased to learn that there are well-preserved wooden lances (or spears) used by archaic humans fully 400,000 years ago. (p. 182-183) I also found interesting his digression on what caused the extinction of the megafauna of America some 10,000 years ago (in Chapter Six, "The Great Extinction").

Primarily, though, this book is about the cultural, behavioral and conceptional abilities (as derived from the archaeological evidence) that separate humans from other living creatures, especially the Neanderthals. Arsuaga reveals his purpose on page 280: "I have been trying to summarize the evidence available concerning the thorniest problem of human evolution, the development of consciousness, which is the defining characteristic of humankind." He had asked in the Prologue on page ix, "Apart from us, has there ever been a life form on earth that was conscious of its own existence and of its place in the world?" In short his answer is yes, the Neanderthal, whom he defines as our contemporary, not as an archaic human species. (p. 278)

Arsuaga's story begins about 2.5 million years ago when *Homo habilis* emerged in Africa (presumably from another post-australopithecine species) with a noticeably bigger brain than the first upright walking apes. "A short while later" (geologically-speaking) *Homo ergaster* (probably the same as *Ho-*

mo erectus) appeared. Arsuaga sees *Homo ergaster* as the first hominid to migrate out of Africa about 1.5 million years ago, spreading to Europe and southeast Asia. Not only did these proto-humans have a significantly larger brain than *Homo habilis*, they had also begun "to create a social and cultural environment...that afforded them ever more independence from the physical environment," which is one of the reasons they were able to survive in diverse climates, especially in the cold of the northern latitudes.

Then about 300,000 years ago Arsuaga sees the development independently in Europe and Africa of a "second great expansion of the human brain" producing "somewhat different results." (p. 307) When modern humans again emerged out of Africa about 150,000 years ago they arrived in Europe to find the Neanderthal. The somewhat different results of their independent evolution prevented the species from merging and eventually the Neanderthal died out.

Although some authorities have emphasized competition with modern humans as the reason for the Neanderthal's demise, Arsuaga believes we need more information before we can say "in a convincing fashion" what happened. (p. 292) He does say somewhat imprecisely that the Neanderthal was "defeated by the cold" while the Cro-Magnons due to "superior technology," especially with bone awls and needles to fashion well-fitting animal skins, etc., were able to survive the glacial maximum 25,000 years ago. (p. 302) However on page 78 while noting that the Cro-Magnons had developed physical features that made them look relatively childlike—a gracile build and a "small, minimally protruding face," ("neoteny" is the technical term for this phenomenon)—Arsuaga may have tipped his hand. He observes, "Cro-Magnons must have looked cute to the Neanderthals! They may have discovered later, to their dismay, what kind of people they were dealing with, and as sweet as the Cro-Magnons may have looked, what kind of behavior they could expect."

(I had a sudden vision here of an abandoned Cro-Magnon child found by a Neanderthal family. They tenderly take the child in, nourish it and bring it up as their own. At a certain age, the child realizes that it is not Neanderthal and... Well, I'm sure there are a few science fiction stories that resolve this premise for better or for worse.)

Arsuaga's account therefore presents the Neanderthal as a co-existing species, not our ancestor, with whom there was little to no interbreeding. Nonetheless Arsuaga has great respect and affection for the Neanderthal. He writes on page 284 that "It would thrill me more than anything if I could say that I had even a drop of Neanderthal blood to connect me with those powerful Europeans of long ago." His portrait is of a "human" species different from (not less than) ourselves that had culture and ritual and self-adornment (as evidenced by, e.g., the Neanderthal's necklace found at Grotte du Renne in France), a being that had achieved consciousness, although of a sort undoubtedly different than ours. In one particular, Arsuaga argues that the fossil evidence suggests that the Neanderthal's phonetic apparatus would not have been able to produce "sounds as distinct as ours," (p. 268). This physical characteristic may have reduced its ability to develop a culture as extensive as the Cro-Magnon's which we know about in part through the cave murals that they painted in France and Spain.

What one feels strongly from Arsuaga's account is the sense of loss that the Neanderthal is no longer with us. How much we could have learned from a being that was at once much like ourselves, but intriguingly different.

Baker, Robin *Sperm Wars: The Science of Sex* (1996) ****
It's a jungle in there

This is a somewhat amazing book on how women can collection sperm from different partners and how the collections of sperm will duke it out within her body in the chase to fertilize the egg, among many other amazing things, including the fact that some sperm serve not to swim after the egg, but to block another man's sperm from the chase. Written from an evolutionary point of view, Robin Baker's text is very readable and certain to make many people uncomfortable. It has had a remarkable effect on me. I suddenly realized how insignificant our consciousness is even in something like reproduction. So much goes on beneath our consciousness, and many things within our consciousness are done for reasons we don't understand or are mistaken about. For example, according to Robin Baker, masturbation serves a reproductive purpose! I won't try to explain here, but he convinced me. Also group sex may actually help a husband to get his wife to bear his child! Read it. I kid you not.

Women come off pretty much as unconscious instruments of the process, men to a slightly lesser degree. All this is as I have always thought, but I had no idea about the details, and I mistakenly thought people, as conscious beings, had a greater effect on reproduction than we actually do. Incidentally at least ten percent of our children are not fathered by the husband, and close to twenty percent of conceptions are from sperm other than that of the husband! (a revelation not unique to this book). "Nowhere is there a woman true and fair," spake the poet. The duplicity of sex is required according to Baker because the woman needs to simultaneously mate with the champion (which is what she is always trying to do) while at the same time keep a man around to help take care of the offspring. Implicit in the book is the idea that people naturally cheat on their spouses as a strategy, a strategy that has consequences, both positive and negative. Sexually speaking, as in everything else, we are instruments of the process more than we think.

This is an excellent, if somewhat creepy, book with the tales of sex and infidelity and scheming by both sexes ringing entirely true. But strange to say I feel like a Victorian, wanting to have a nice cup of tea and talk about something else.

Calvin, William H. *A Brief History of the Mind* (2004) *****
The rise of "beyond the apes" intelligence

The central event in this book is the human mind's so-called "big bang" which occurred some 90,000 to 50,000 years ago.

(These are neurobiologist William Calvin's numbers from page 111 where he notes that "it now appears that humans were behaviorally modern before the last great Out of Africa" which is now understood as taking place between 60,000 and 40,000 years ago as determined by the latest tweaking of the mitochondrial DNA dating data.)

Professor Calvin leads up to this event by starting with the proto ape that was our ancestor (and the ancestor of modern apes) that lived some seven million years ago. He takes us from that ape's jungle habitat to the woodlands, where our ancestors learned to walk upright, and from there to the savannas where they ran down, killed and ate large game animals. Somewhere along the way we got smart. But, Calvin wonders, did we get smart enough?

He sees a disconnect between our abilities and the world we have inherited. He asks: "Where does mind go from here, its powers extended by science-enhanced education and new tools—but with its slowly evolving gut instincts still firmly anchored to the ice ages?" Are we just a "rough-around-the-edges prototype, the preliminary version that evolution never got a chance to further improve before the worldwide distribution occurred?" (p. 178)

In other words, are we using Stone Age instincts to cope with Information Age problems? It is interesting to note that in psychologist Keith Stanovich's recent book *The Robot's Rebellion: Finding Meaning in the Age of Darwin* (2004) he is concerned with the same problem from an entirely different point of view. He writes about the "potential mismatches between the cognitive requirements of the Environment of Evolutionary Adaptation and those of the modern world."

Of course the problem, as both writers point out, is that cultural evolution out-sprints the biological so that our genotypes are still in the woodlands and on the savannas as the ice ages come and go while our phenotypes have to deal with traffic jams, weapons of mass destruction, and the paperwork for our HMOs.

One of Calvin's more intriguing ideas comes from his dictum that "behavior invents and...New form follows new function." (p. 159) He argues that the higher intellectual functions of humans came from the development of a "structured suite" of brain machinery that "is shared in part with some nonintellectual functions." (p. 94) He sees "accurate throwing" as part of this structured suite, and argues that learning to hit a moving target (say a small animal), because it involved several parts of the body (hand, wrist, arm, shoulder—and eyes and legs for that matter) in close coordination, several parts of the brain were used simultaneously as well. Consequently a "structured suite" developed in the brain that later was used for the development of symbolic language. What he is saying is that the syntax of language—that is, the "something" does "something" to "something": the subject-verb-object structure of language that works so magically for us—actually came from the body's experience running down game in Africa.

I think Calvin is on to something here because that syntactic structure which is common to people everywhere, regardless of what language they speak, mirrors the action of the world. What is important in the environment is what is being done or what is happening (the verb), by whom and to whom (or what): the bull gores the lion; the monkey peels the fruit, the wind blows the tree down, etc.

Another of Calvin's pet ideas is that education "perhaps more than any of the imagined genetic changes" is what will best help us cope with the challenges of the modern world. (p. 184) He argues that if children are exposed to "structured stuff" at younger ages, and if they can "softwire their brains to better handle" such stuff, "the more precocious children will soon double the amount of structured speech heard by the next generation." (p. 154)

Of course our brains are still being "softwired" after we leave the womb and for some many months afterwards as our experiences serve to strengthen certain neurons and discard others. It seems, however, that Calvin is getting at something larger here, a kind of quasi-Lamarckian accelerated evolutionary process. Indeed I think he intends this example as a possible explanation for the "big bang" that took place in the Pleistocene. To be honest I have no idea whether he is right or not. Certainly it is an interesting idea.

Interesting is this comment from page 104: "[M]uch of [our] higher intellectual function seems half-baked, what you ordinarily see in a prototype rather than a finished, well-engineered product. Perfection you don't get, not from Darwinian evolution...But culture...can sometimes patch things up, if society works hard enough."

This is my first experience with reading Calvin, and I can say that reading this book is like engaging in a conversation with a wise and learned man who likes to share his ideas.

Diamond, Jared *The Third Chimpanzee: The Evolution and Future of the Human Animal* (1992) *****
Brilliant with some flaws

After the spectacular success of UCLA geography Professor Jared Diamond's *Guns, Germs, and Steel: The Fates of Human Societies.* (1997) and *Collapse:*

How Societies Choose to Fail or to Succeed (2005), Harper Perennial has reissued this splendid book originally published in 1992. I'm glad they did. In this ambitious work, Diamond attempts to define human nature in evolutionary terms and to warn us of the dangers ahead. He is particularly worried about the two clouds he sees as hanging over our heads, nuclear warfare and an environmental holocaust. (p. 350)

In the early chapters, Diamond examines our nature and shows how we are similar to and differ from the common chimpanzee and the bonobo (AKA "the pygmy chimp" or as lately seen on TV, "the sexy chimp"). His expertise is nothing less than stunning. Even though this book is nearly 15 years old, most of what he writes needs no update. In the later chapters he concentrates on a variety of themes, genocide, the "noble savage," environmental loss, species extinction, etc. Here we can see the tentative ideas that later became the books mentioned above.

In chapter one, Diamond compares the three chimps. In chapter two he documents the so-called "great leap forward" about 40,000 years ago in which humans became truly human as evidenced by cave art, better tools and the ability to improve upon previous tool design. He attributes this leap to the development of symbolic language. In chapters three through six, he examines human sexuality and reproduction. In chapter seven he explains why we grow old and die. Chapters eight and nine explore language and art and their expression in other animals. In chapter ten, "Agriculture's Mixed Blessings," one of the best chapters in the book, Diamond shows us that life as a hunter-gatherer was preferable to life as one of the early agriculturists. With agriculture came the possibility of civilization and everything that civilization brings, which includes—in addition to art, technology and the massive harnessing of energy—herd diseases, malnutrition from monoculture farming, overpopulation, and hard and long work hours for most people. Average human height actually decreased following the birth of agriculture about 10,000 years ago.

In chapter eleven Diamond begins to stray from what he really knows to what he thinks he knows. He posits here that we drink and use "dangerous drugs" because of a macho need to show how fit we are. He takes Amotz Zahavi's famous handicap principle and applies it to the Marlboro man. But the advertising for cigarettes and alcohol that Diamond sees as appealing to fitness are better seen as appealing to a false sense of glamour or adventure. Actually we use drugs because they alter our consciousness or deaden it; and we continue to use them because we become psychologically dependent on them. A way of looking at drug use that is consistent with evolutionary principles is to see drug use as a relationship between species, between plant (producer of, e.g., nicotine, tetrahydrocannibinol) and human, or between yeast (alcohol) and human which has not yet reached a true symbiosis.

Another error that I think Diamond makes is his idea that intelligent species throughout the universe are unlikely. He uses the argument that intelligence arose only once on this planet and that if it was something that evolution could easily develop it would have arisen in other species, but hasn't. He even recalls an analogy that I've read elsewhere from woodpeckers. Noting that there are no native woodpeckers in Australia, it is postulated that although woodpecking is a fine subsistence niche, it requires such exacting skills that its evolution almost didn't happen on this planet. The same may be said for intelligence as an evolutionary skill. But the fact that woodpeckers already exist in the Americas and the Old World tends to preclude the evolution of other birds into woodpeckers. And who's to say what intelligent creatures might have evolved had we not come along (e.g., the Neanderthal)? And who's to say just how intelligent some dinosaurs were before they were wiped out? And who's to say what a colony of ants or bees ("swarm intelligence") may become after we are gone?

I also think Diamond is missing something when he declares that "advanced extraterrestrials who discovered us would surely treat us in the same" barbaric way we have treated other primates (or indeed other peoples). (p. 214) For one thing those little green men, considering the vastness of interstellar space, would have a hard time getting here, and any that did arrive here would be light years in advance of us not only technologically but probably morally as well.

A recurring theme throughout the book is the human propensity to kill and our hypocrisy about that killing. From the mastodons to the children of the Middle East, humans have always killed while maintaining that killing is evil. Diamond does a nice job of explaining just how this Orwellian doublethink works. The main mental trick is to see those we want to kill as different and separate from ourselves. The taboo against killing humans, Diamond reveals, is really just a taboo against killing members of our own family and tribe. Once we are able to see others as outsiders, we can demonize them and trivialize them, turn them into subhuman objects and get on with the slaughter. Diamond considers how those of us on the sidelines, those of us who have not demonized the victims, can let this happen. His conclusion is that human nature can stand only so much blood-splattered horror before we become numb to the killing and turn away.

Although Diamond waxes hopeful near the end of the book as he thinks of his children and grandchildren, the overall impression I got is that humans are probably not going to be able to prevent the twin nuclear and environmental holocausts to come.

Diamond, Jared *Why Is Sex Fun? The Evolution of Human Sexuality* (1997)
****
The least and shortest of Diamond's books, but excellent nonetheless

Sex is urgent, demanding, sometimes pleasurable, but fun? No, I would not call sex fun. By calling sex fun I think Professor Diamond skips over the very essence of sex which is *we have no choice*. That's the way it has come down to us. As one of my students crudely put it, "eat cheese or die."

Diamond knows this of course as do all of us. What he is about in this his second book, coming after *The Third Chimpanzee* (1992) and before the phenomenally successful and highly recommended *Guns, Germs, and Steel* (1997) is to show the general reader how evolutionary biology and the study of the sexual behavior of other species can focus light on human sexuality. He considers such questions as why our sex life is the way it is and how such behaviors are adaptive. He goes beyond the well-known phenomenon that men seek a lot of one night stands while women look for men with resources and a willingness to commit to a long term relationship. He even goes so far as to speculate on why men don't breast feed, while intimating that evolutionarily speaking it might have been something that would work for the human species.

He looks at the battle of the sexes from a strategic and a biologically adaptive point of view. He uses studies from primatologists, anthropologists and others, as well as his own experiences in New Guinea where he studied birds and came to know well the indigenous people and their habits. He offers insight into why women in most societies end up doing most of the work while men are out "big game" hunting and playing "show off." He shows how the occasional large kill is more about status within the tribe than it is about nutrition. He makes it clear that the key to understanding the division of labor between the sexes depends largely on how much nurturing is required before offspring can take care of themselves. By studying birds, whose reproductive behavior vis-à-vis monogamy is most similar to humans, we can see that parental demands are usually too great for a single parent. Therefore both birds and humans are more or less monogamous. This is in contrast to our biologically closer cousins, the overwhelming majority of mammals, who are raised almost exclusively by the mother while the father is busy looking for the next reproductive try.

Interesting is how the reproductive strategies of chimps, gorillas, orangutans and humans differ. Orangutans live solitary lives and meet briefly to mate while chimpanzees are mostly promiscuous, especially bonobos, who even more than humans, use sex as a means of social bonding. Diamond presents theories on why ovulation in humans is concealed even from the woman herself and why this has proven effective in an evolutionary sense. (Either the male must stay home and guard his mate continuously since he doesn't know when she's fertile, and/or the woman must "trick" the males into thinking that anyone of them who had sex with her might be the father and therefore none of those males is likely to harm her baby.)

What I found most enlightening was the answer to the question (incidentally not asked in this book) why are so many human societies and religions patriarchal? The answer, it dawned on me while reading the chapter entitled "What Are Men Good For?" is that patriarchy is a strategy by males to counter the uncertainty posed by the hidden ovulation of women! It's all part of the battle of the sexes. Woman gained control with hidden ovulation since they would always know who the mother was, but men would be in doubt about who the father was. Enter social and political control of women so that paternity is more nearly certain.

In the chapter "Making More by Making Less" Diamond explains why women experience menopause and men don't and why it is almost absent in other animal species—the pilot whale being a notable exception. It seems that a woman getting on in years can better ensure the success of her genes by using her energies and her hard-earned knowledge to help rear her grandchildren instead of getting pregnant again. This idea is closely aligned with ideas about senescence. We get old and die because our systems run down and/or suffer accidents. They run down because the evolutionary mechanism doesn't "care" about people past the reproductive age (natural selection no longer works on non-reproducing life forms!). But why should our reproductive abilities end while we go on living? Because, due to the rigors of life in the wild, older people are not as capable physically as their children and other young people, and so they get selected against. Diamond even goes so far as to intimate that grandmothers by forgoing having more children benefit the entire tribe with their efforts at foraging and being a repository of knowledge about what happened long ago and how to survive rare catastrophic events. As for men, well, their reproductive abilities run down more slowly, but after a certain age it is all the same.

Klein, Richard G. with **Blake Edgar** *The Dawn of Human Culture: A Bold New Theory on What Sparked the "Big Bang" of Human Consciousness* (2002) *****
Neither "bold" nor "new," but excellent nonetheless

Professor Klein and science editor Blake Edgar refer to "innovation" as the key to the great leap forward made by humans about 50,000 years ago. This was the beginning of human culture—the "dawn" as they call it. It wasn't a change in physiology—humans have been anatomically modern for something like 150,000 years. What changed was the wiring in the brain, or the chemistry in the brain or the linkage between the modules in the brain, or, as they express it, there was a "neurological shift"—at any rate, something that would never show up in a fossil.

This is Klein's theory and it is a persuasive one, albeit one that can never be proven—well, probably can never be proven. If under some ice sheet (as the planet continues to warm) we find a 100,000-year-old human intact, perhaps an examination of his or her brain and a comparison with the modern brain will give us the proof. Barring that very unlikely event, there is no way we can see what changed.

But it doesn't matter. Formal proof of Klein's conjecture (and of course, he is hardly the first to present such a theory) is unnecessary. We know from the behavioral changes that took place in something like a twinkling of an eye that humans beginning about 50,000 years ago were suddenly different. They had a culture that developed from the use of what might broadly be called symbolism. We can see this in the petroglyphs and cave art and artifacts that they left. We can also see it in the way they displaced the Neanderthals in Europe and left no trace of *Homo erectus* elsewhere in the world, and how quickly they spread to the far corners of the planet.

It is easy to see that they must have had symbolic language as well. Indeed, I think language really is the key to what happened, and this is Klein's point as well. The key idea is that "language is almost a kind of sixth sense, since it allows people to supplement their five primary senses with information drawn from the primary senses of others." (p. 146)

Today's mighty culture would be impossible without written language or some means of taking down and recording and maintaining human knowledge. Prior to written language this was done through an oral tradition handed down from one generation to the next. Myths, stories, poetry, ideas, information and methods were memorized and recited. Prior to that however, prior to the use of symbolic language, there would have been only a limited ability to pass ideas down from one generation to the next. It would have been difficult to even share some ideas with a contemporary. But once symbolic language developed, people could demonstrate events and things not present with others through the use of words—that is, symbols standing for the actual objects or events—nouns and verbs.

From a representation symbolically of something seen or something that happened, it was only a step to a representation of something never seen before—such as a net for catching birds or fish or a stampede of wildebeests over a cliff.

This is the innovation that Klein refers to. This is the difference between the Late Stone Age culture and the Middle Stone Age culture, between the Upper Paleolithic and the Mousterian. A human arm can throw a spear, but a human arm extended with a lance can throw the spear farther and with more force. People could travel only so far without water, but a people who carried water in skins or watertight baskets (not preserved in the fossil record obviously!)

could travel much farther. Actually I imagine that the first truly modern humans may have carried soup—yes, soup with its sterile, boiled water—in skins on their backs!

What this book is about then is a close and detailed description of the progression from archaic humans to fully modern humans. It is a carefully constructed argument that shows that the change was not gradual, as some would have it, but abrupt. Whatever one may think about Gould and Eldredge's punctuated equilibrium, Klein makes it clear that in the case of human evolution, a key transformation—indeed THE key transformation—occurred quickly. The most persuasive part of their argument is that the "new" humans were able to not only dazzle us with their symbolic art, etc., they were able to grow their populations and thrive in places where humanoids had never survived before.

This book is also full of interesting information about archeology and anthropology, including how fossils are dated and theories developed. One of my favorite tidbits is this: the size of archaic human populations could be surmised by the size of tortoise bones! Since tortoises were relatively easy to catch, the biggest ones, "the most visible and the most meaty" would have been taken first. So as "the number of collectors increased, average tortoise size declined." (p. 166)

For many readers, the most interesting part of the book might be the distinction that Klein and Edgar make between *Homo neanderthalensis* and *Homo sapiens*: "It doesn't follow that Neanderthals and modern humans couldn't interbreed or that they never did, but the DNA results strongly support fossil and archeological findings that if interbreeding occurred, it was rare...this inference, together with fossil evidence...justifies their assignment to...separate species..." (pp. 185-186)

This is not an easy book, but it is not unnecessarily difficult either. I think Klein and Edgar did a good job of treading that fine line between being too technical (and jargony) and not technical enough.

By the way, despite the sensational subtitle, the authors scrupulously and wisely avoid using the word "consciousness" throughout, and nowhere do they speak of a "Big Bang of Human Consciousness."

Potts, Malcolm and **Thomas Hayden** *Sex and War: How Biology Explains Warfare and Terrorism and Offers a Path to a Safer World* (2008) *****
One of the most important books I've read in years

Potts' main thesis is that all men have the potential to kill other people to get what they want or because they are told to kill or because they have dehumanized their victims. All men—you, me, and Professor Potts himself, but for the grace of God, could be in Darfur slicing people up with machetes. All that is required is that the victims be seen as members of an outgroup as opposed to the ingroup to which we belong.

This is a startling thesis, one that sets the standard social science model, in which it is said we have to be carefully taught to kill, on its head. What Potts says is that the violence we have seen throughout human history is innate, an evolved trait that was once useful for hominids in the tribal setting. This is also the thesis of evolutionary psychology. Instead of learning to kill, or being taught to kill, we need to be taught NOT to kill. We don't usually kill members of our family or friends because they are part of an ingroup with which we identify.

Potts has a solution, which is why he has written this fascinating and exhaustive treatise on war and its causes. His solution begins with an understanding that our psyches are governed by evolved Stone Age emotions similar to what we see in chimpanzees as they conduct their horrific raids on isolated individuals from neighboring groups, ripping and tearing their victims apart with their bare hands and teeth. Potts calls this "team aggression," a strategy that has been perfected in human beings. Men bond together and use their greater numbers to kill members of other tribes so as to gain resources such as territory, slaves and women to impregnate.

In the modern world we have men with Stone Age brains in positions of power with their fingers on weapons of mass destruction. We know that they will posture and threaten and eventually convince themselves of the evil of the enemy and pull the trigger.

Understanding all this, Potts moves to the solution. Since it is men—not women—who engage in team aggression, we need to put women in positions of power since they have proven to be less likely to go on killing raids. (Potts presents a formidable amount of evidence to support this idea.) Furthermore, the average woman needs to be empowered to the extent that she can choose when and if to have children. Potts shows that countries with large and growing populations relative to resources are more likely to engage in raids on their neighbors than countries with stable populations. Additionally, it is the demographic makeup of the population that is significant. A country with a large percentage of young men relative to older men and women tends to be more violent. Women in sub-Saharan Africa for example typically do not have access to contraception and family planning. Consequently they (and women in the Middle East as well) typically have six, seven or eight children in their lifetimes. Rapid population growth is the result which strains resources and leads to a society with a lot of young men in it who have little to lose and so are easily led to acts of violence.

He adds: "Fundamentalist teachings, whether Christian, Muslim, or any other religion, end up restricting and controlling women, which in turn makes wars and terrorism more likely. The twenty-first century is seeing a clash of cultures, but that clash is not between Islam and Christendom. Rather it is between fundamentalism and reason." (p. 363)

Potts notes that "In the past fifty years the world has accommodated rapid population growth tolerably well, although as rising oil and food prices suggest, this may not be true in the future." He compares us to the "first people to cross into North America, or the Polynesians who first landed at Easter Island...Presented with vast new supplies of food, energy, building materials, and luxury goods our forebears could never have imagined, we have gorged ourselves on consumption, and we have driven our global population...to six billion in 2000... The evidence of that increase is now all around us, in our polluted environment, our warming climate, our disappearing rainforests, and our increasingly degraded farmland: We are, as a species, in the process of proving Malthus's proposition that population will always outstrip resources." (pp. 296-297)

We are Easter Island natives. We have arrived not at an unspoiled island with flightless birds and a virgin forest to ravage, but at a planet with resources still rich enough to exploit and a powerful science and technology to do the exploiting. It took a few hundred years for the Easter Islanders to deplete their resources and return to a mean and savage, poverty-stricken existence. How long will it take us?

Potts writes, "...it is highly likely that our numbers and industrial demands have already exceeded the environment's capacity to support them. Mathias Wackernagel in California, Norman Myers in England, and others calculate that we may have exceeded Earth's carrying capacity as long ago as 1975. According to these calculations, we already need a planet 20 percent larger than the one we have." (p. 299)

There are two points that Potts does not dwell on that I want to emphasize. First, wars have the ability to fix the problem of too many young men with nothing to do. Second, women make sexual choices and in doing so often choose the most violent men to mate with because they know that such men are more likely to survive in violent times and provide for their children than less violent men. Women in precarious situations do not make moral judgments. Instead they make realistic ones.

Wade, Nicholas *Before the Dawn: Recovering the Lost History of Our Ancestors* (2007) *****
How DNA analysis is illuminating the prehistory

I thought the first part of the book which was actually about the prehistory as newly discovered through DNA analysis was very interesting. I was less thrilled with the chapters on Race, Language and History. The wrap up chapter on Evolution was good, if a bit repetitious.

Wade writes extremely well and does a good job of summarizing the latest (circa 2005) research, much of which has come from analyses of the descent of the Y chromosome (from men) and mitochondrial DNA handed down through the female line. The question of our relationship with the Neanderthal—long a thorny question—is more or less resolved with DNA extracted from Neanderthal fossil bones that has been compared to the sequences of human DNA. The conclusion is that *H. neanderthalensis* came from *H. ergaster* through *H. heidelbergensis* as *H. sapiens* did, and then broke off on its own.

Furthermore there is no genetic evidence that human and Neanderthal produced viable offspring. The earlier idea than the Neanderthal was a modification of the very successful *H. erectus* has been discredited.

As to the question of our origins, northeast sub-Saharan Africa is further confirmed as the site. Wade has humans becoming behavioral human around 50,000 years ago after becoming anatomically human as early as perhaps 200,000 years ago. The great leap forward occurring 50,000 years ago is attributed to the acquisition of symbolic, syntactic language. This was also the time when humans made the exodus out of Africa and began to colonize the world. They went east across the Red Sea at the Gate of Grief during a glacial period when the sea level was two hundred feet lower than it is today. They followed the coast line of the present Gulf of Aden and the Arabian Sea to India and eventually to Australia. I had previously though humans had gone north along the Red Sea to the Mediterranean and then east and then north to Europe. However, the evidence indicates that it was only later that humans migrated to Europe from India westward to replace the Neanderthal.

I had also always thought that agriculture came before settled communities, but it now appears that sedentism occurred first and was part of a behavioral and psychological change in humans that led to agriculture and eventually to cities and nation states. Just prior to or at about the same time as the first settlements appeared some 15,000 years ago occurred the domestication of the dog. Wade avers that living in settlements near a plentiful food source (wild grains, a bountiful river, etc.) was partially made possible by people using dogs as sentries against the ancient practice of dawn raids by neighboring tribes. Clearly the transition from the hunter-gatherer way of life to the settled way of life was a momentous one.

Perhaps the reason I wasn't so thrilled with the latter part of the book is that I read some of the studies Wade considers elsewhere. The experience of Brian Sykes in tracing the ancestry of people named "Sykes" and of Thomas Jefferson's second family with the slave Sally Hemings are examples of DNA derived stories that I had read before. Wade's account of the saga of the Ashkenazi Jews of Europe, although also a familiar story, is most interesting. He cites studies showing that Ashkenazi Jews have an average IQ of 115 while Sephardic and Oriental Jews have the usual average of 100. A couple of arguments are presented to account for this difference. The more plausible one is that because the Jews of Europe were forced by the Christian majority into becoming money lenders from about AD 1100 until around 1700. (Christianity at the time forbade usury.) That sort of intellectually demanding way of life, along with having to make a living amid persecution, selected for intelligence. By way of contrast, Sephardic and Oriental Jews during the same period "lived mostly under Muslim rulers who often forced them into menial jobs, not the intellect-demanding ones imposed on Ashkenazim." (p. 256)

More than any other book I have read, "Before the Dawn" insists on cultural change leading rapidly to genetic change. With the experience of the Ashkenazi Jews as a case in point, Wade argues more generally that "for social species the most important feature of the environment [which directs evolutionary change] is their own society." He concludes that "to the extent that people have shaped their own society, they have determined the conditions of their own evolution." (p. 267.

This might be termed "evolution by your own boot straps." I wonder however if it isn't a sort of fallacy. Biological evolution shapes human behavior which in turn leads to cultural change which leads to further biological evolution. I think it is better to speak of cultural evolution as a subset of biological evolution and not imply that somehow we have begun to direct the process. But this may be just a quibbling over semantics. Clearly the environment has changed us and we have changed the environment.

In the final chapter Wade speculates on where we are going. I always like such speculations but only really appreciate those that have us becoming post-human in some way. Wade posits one possibility that I have not thought about in years, that of humans splitting into two or more species. He notes: "Our previous reaction to kindred species was to exterminate them, but we have mellowed a lot in the last 50,000 years." (p. 279)

By the way, this idea that we "have mellowed a lot," and become less aggressive since we have domesticated ourselves is one that appears elsewhere in the book and is an idea that, for better or for worse, appears surprisingly to be true. The actually percentage of humans killed during warfare appears to have been much greater during the prehistory than it is today. The wars today are much bigger but the wars in the pre-history, according to the research presented here, were nearly constant.

Zimmer, Carl *Smithsonian Intimate Guide to Human Origins* (2005) *****
Engaging text, beautiful artwork

The romantic days of the search for the "missing link" are gone, and as science writer Carl Zimmer reminds us, that is all to the good since the very idea of a "missing link" is a misdirection. What we have today is the search for human ancestors and for a distinction to be made between our ancestors and other ancient hominids. This book with its beautiful prints and photos, engaging drawings and helpful charts, and especially the sprightly text by Zimmer brings the general reader up to date (circa 2005) on the latest developments.

There's a lot going on. There's the controversy about *Homo floresiensis*, thought to be a tiny hominid, found in Indonesia in 2004. Zimmer presents the arguments. Some think that *Homo floresiensis* is an island adaptation of *Home erectus*, the first hominid to make it out of Africa 1.8 million years ago. After all, island adaptation often leads to diminished size. There are fossils of now extinct small elephants in Indonesia. But others believe that the skull found is an anomaly, a case of microcephaly, a birth defect. I'm betting on the latter. [Note: since I wrote this I've changed my mind. From the latest evidence that I am aware of *Homo floresiensis is* (probably) an island adaptation of (possibly) *Home erectus*.]

There are wooden spears found that are around 400,000 years old, meaning that *Homo habilis* or *Homo ergaster* (who may be one and the same) or the more recently discovered *Homo heidelbergensis* were accomplished tool makers long before *Homo sapiens* arrived on the scene. There is the idea that *Homo neanderthalensis* is a cold climate, European adaptation of *Homo erectus*. [Note: more recent evidence suggests that *Homo neanderthalensis* is later than *Homo erectus* and more closely related to us through *Homo heidelbergensis*.]

Part of the excitement in paleontology is in the new fossil finds, and part is in our new-found ability to analyze DNA samples to map the spread of hominids. This allows us to see the "out of Africa" phenomenon in three main stages: (1) *Homo erectus* leaving Africa 1.8 million years ago, followed by (2) *Homo heidelbergensis* expanding into not only Europe and the Near East and China, but into Southeast Asia as well. Finally (3), about 130,000 years ago, *Homo sapiens* begin to move out of Africa, first into the Levant and then into East Asia and Australia (50,000 years ago), then into Europe and Siberia (40,000 years ago) and ultimately into the Americas (20,000 years ago). Incidentally, this book has *Homo sapiens* coming onto the scene almost 200,000 years ago.

Zimmer talks about the various hominid cultures and speculates on their social and religious possibilities. On the subject of what happened to the Neanderthal, he intimates that he believes it was a combination of things that allowed humans to survive while the Neanderthals went extinct, including being better able to adapt to climate change, having a more sophisticated culture and better hunting techniques. I think it's also possible (actually I think it's likely) that humans were better at killing not only herd animals but the competition as well, meaning that one of the reasons that the Neanderthals are gone is because we killed them. Zimmer more or less skirts around this, waiting (wisely, I think) until further evidence is in.

In a final chapter, "Where Do We Go from Here?" Zimmer briefly discusses biotechnology and genetic engineering, and how our species might be affected by cultural evolution.

This is a handsome book. It's like a coffee table book with the high gloss, heavy pages and the beautiful artwork, but smaller in size. Most significantly it is a book aimed at the general reader that is well written, well edited, and very well presented. And it is clear. It is in fact the clearest book on human origins—usually a very murky subject—that I have read.

By the way, Zimmer is the author of several excellent science books. I especially recommend his creepy, but fascinating, *Parasite Rex: Inside the Bizarre World of Nature's Most Dangerous Creatures* (2000).

Chapter Two

Evolution vs. Creationism

Some of the books reviewed in this chapter pretty much destroy the notion of creationism or its latest tuxedo impersonation, "Intelligent Design" AKA "Unintelligent Design." I have included reviews of books by authors who are not biologists because I think they shed some light on the subject. Two of the authors speak from a Christian perspective (Caiazza and Dowd) but come to differing conclusions. Another (Hunter) spuriously argues the creationist case, but not, in my opinion, to very good effect. One of the authors, Edward Humes, is a truth crime journalist who covered the Kitzmiller v. Dover "second Scopes trial" in Pennsylvania in 2005. Included is a book by famed biologist Edward O. Wilson who tries to make nice with creationists while gently pointing out the paucity of their fraudulent position.

Ayala, Francisco J. *Darwin and Intelligent Design* (2006) ****
Reconciling biological evolution and religion without Intelligent Design

Judging from the modest length of this book (a little over a hundred pages) and from the clear, straightforward prose, and from the fact that it is part of the Facets series of the Fortress Press, I would say that the purpose is to inform public opinion at the most elementary level. Since the Fortress Press is "the ministry of publishing for the Evangelical Lutheran Church in America" one can expect a Christian perspective. What is interesting is that that perspective is clearly in acknowledgment of the truth of evolution and the non-scientific nature of Intelligent Design.

The author, Francisco J. Ayala, who is a professor of both biology and philosophy at the University of California, Irvine, presents the ideas and arguments in a way that even junior high school students can understand. He is fair and he is unmistakable. He argues that science and religious beliefs need not be in contradiction. He believes that the way to achieve this understanding is to avoid reading the Bible or other religious works in a literal sense, and leave the science to the scientists and the scientific method. This is the position of most educated people that I know of, a position advanced by both religious leaders and scientists. His purpose, as he expresses it in the Prologue, is "to convince" the reader "that we may accept" the scientific evidence for the truth of biological evolution "without denying the existence of God or God's presence in the universe..." (p. vii)

In short this book represents a "middle of the road" position in the evolution versus ID debate. Richard Dawkins, et al., are not going to be convinced that one can believe in a personal God who watches over us while at the same time acknowledging the truth of Darwinian evolution. Similarly, most conservative and evangelical Christians will not be pleased since the text rejects a literal interpretation of the Bible while favoring a seemingly purposeless evolutionary mechanism.

Personally, I greatly favor Ayala's position and approach. I think it is essential that we understand that science need not impinge on matters of faith; and that religion should not pretend to scientific truth. Science is a method for achieving a better understanding of our world and how it works, and as a guide to the development of ways to better manage our environment to our advantage. Religion is a method for guiding us toward an understanding and appreciation of questions that cannot be addressed by science.

Brockman, John, ed. *Intelligent Thought: Science Versus the Intelligent Design Movement* (2006) *****
Amounts to a destruction of Intelligent Design

As Editor John Brockman writes in his introduction, this book, a collection of 16 essays by eminent scientists, "is a thoughtful response to the bizarre claims made by the ID movement's advocates, whose only interest in science appears to be to replace it with beliefs consistent with those of the Middle Ages." (p. x)

What the ID people are about is a power grab, an attempt to install themselves as The Authority on who we are and how we got that way. God is the puppet for whom they speak. As Brockman further notes, theirs "is a duplicitous public-relations campaign funded by Christian fundamentalist interests." (p. x)

Following the original and very interesting essays by Daniel Dennett, Richard Dawkins, Steven Pinker, Lee Smolin, Stuart Kauffman and eleven others is an incisive excerpt from the "Memorandum Opinion of the United States District Court for the Middle District of Pennsylvania" in the case of Kitzmiller v. the Dover Area School District, dated December 20, 2005. Judge John E. Jones III, in ruling for the Plaintiffs, makes it abundantly clear that ID is not science and has no business being taught in science classes. He chastised some members of the Dover School Board (who have since been voted out of office), noting that "It is ironic that several of these individuals, who so staunchly and proudly touted their religious convictions in public, would time and again lie to cover their tracks and disguise the real purpose behind the ID Policy." (p. 254)

Dawkins, whose essay is entitled, "Intelligent Aliens" has warned us before about the dishonesty of creationists and ID proponents. One might ask, why are they so dishonest? Why do they bully and misrepresent? One suspects they think they have license since theirs is the work of God. At least, if you tell yourself that, as suicide bombers do, and you believe it, then whatever means you use are justified. Which is the reason that it is a waste of time to argue with ID people. They already have the truth and any argument is totally beside the point. They pretend to some spurious debate only for propaganda purposes.

Brockman knows all of this and instead of getting involved in a phony "debate" with the "intelligent design cabal" (Dawkins' designation) what he has done is persuade these sixteen distinguished scientists to explain from various disciplines (philosophy, psychology, biology, paleontology, ecology, even physics) just why, as Theodonsius Dobzhansky so succinctly put it, "Nothing in biology makes sense except in the light of evolution." And they do a great job of that. Additionally, the essays offer insight into the evidence for evolution and further our understanding. Some excerpts:

"A denial of evolution—however motivated—is a denial of evidence, a retreat from reason to ignorance." (p. 80) —paleontologist Tim D. White

"An understanding of morality is to be found through secular moral reasoning and lies in fundamental facts about the human condition, not in the dictates of a supernatural deity." (p. 143) —cognitive psychologist Steven Pinker

This is the point of Pinker's essay, a refutation of the religious idea that human beings cannot be moral without the fear of retribution from God, or that religion is what teaches morality. His striking and very persuasive argument includes the idea that, "an evolutionary understanding of the human condition, far from being incompatible with a moral sense, can explain why we have one." (p. 152)

One of the delusive ideas of the ID people is the notion of "irreducible complexity." The problem with that, as Dawkins has observed, is, how can we be sure that something is irreducibly complex? Physicist Seth Lloyd's essay "How Smart Is the Universe?" demonstrates that the universe is plenty smart enough to handle any sort of "irreducible complexity" on its own without any help from supernatural beings. He notes, "Because of the universe's information-processing power and diversity, it was virtually certain to hit upon life sooner or later." (p. 187)

If you haven't encountered this line of reasoning before—the universe as an information processing computer—(and I hadn't) reading this essay should be most interesting. Lloyd estimates that the universe has performed around 10 to the 122th operations in its 13.8 billion years of existence. (p. 180) Add this computing power (call it the ability to perform trial and error experiments at random) to the self-organizing aspects of matter and energy (as presented in Stuart A. Kauffman's essay, "Intelligent Design, Science or Not?") and the appearance of life in the universe seems well-nigh inevitable—which I believe is the majority opinion of scientists today. Kauffman believes it would contribute to a better understanding if evolution were "recast as a marriage of self-organization and selection." (p. 177) I think this is already being done.

By the way, Kauffman shows how the ID people could make a testable prediction (although, of course, they dare not). He writes, "The intelligent-design advocate must predict that in NO CASE will...intermediate forms [of life] with diverse functionalities be found." His point is that intermediate forms are "evidence against irreducible complexity demanding a Designer." (p. 173) His conclusion is that such forms exist and "count as disconfirming evidence" not pleasing to ID "scientists."

It is interesting to note how the essays and their arguments from diverse fields support one another and amount to unified support for the fact of evolution. This is the strength of the book, brilliantly conceived and nicely put together by John Brockman who is a science editor par excellence.

Caiazza, John C. *The War of the Jesus and Darwin Fishes* (2007) ***
Or the war among those fishes and postmodernism

It should be noted at the outset that Caiazza is Adjunct Professor of Philosophy at Rivier College in New Hampshire which is a Roman Catholic institution. Caiazza's views throughout are consistent with that of the Roman Catholic Church.

The book is not, as the title might indicate, a popular and easily assimilated book. Instead it is a highly intellectual exercise, the main purpose of which is not so much to champion the side of the Jesus fishes as it is to diminish the postmodern world view. Caiazza makes this clear in the final chapter in which he writes, "...without the repairs offered by religion and science it is possible to predict that postmodern culture will continue its descent into intellectual confusion and moral chaos." (p. 164)

Caiazza's technique is to give a historical perspective on the arguments and the modern opinion from both sides of the issue. His bias is clear: he is against reductionism (a term that is often a euphemism for the method of science itself) and sympathetic to not only contemporary religious views, but to what Francis Crick called the mysterian position. Like the creationists and others Caiazza is afraid of losing the "mystery" of human consciousness to a quasi-deterministic organic universe. He makes much of the idea that the indeterminacy of quantum mechanics has killed the determinism of the 19th century. From that he leaps to the conclusion that God has been resurrected and is alive and well in the hearts of physicists as well as in the hearts of Jesus freaks and sober Episcopalians.

What he does so very well is present the assumptions of the postmodern understanding while delineating how we got to such a place. The first chapter "Religion and Science in the Postmodern World" serves nicely as an introduction to postmodern thought. Caiazza is also good at critiquing the stalwarts of science from string theorists to evolutionary biologists while reminding us that scientific facts and theories are always subject to falsification.

Much of the book is, alas, an extended rant against reductionism in science, a view shared by postmodernists. The problem with reductionism is not so much that it is intellectually bankrupt, as reading this book and others might suggest, but rather that it is not rigorously defined. If one means by reductionism that something cannot be greater (in some sense) than its parts, then that reductionism is mistaken. If instead it is meant that typically a thing cannot be understood by a minute examination of its parts, that too is mistaken. Some things are greater than an examination of their parts would seem to indicate, the human brain for example; and some things CAN be understood by an examination of their parts, such as wrist watches and tinker toys. The real problem with reductionism is that it is limited because human abilities are limited. It is a technique doomed to failure when things get very complicated. We cannot, as in Edward O. Wilson's dream of "Consilience," actually trace the steps from atomic particles to human behavior, even though the steps might be there.

Now to some quibbles:

Caiazza writes that "...after three centuries of discovery and application science now presents us with as many problems as solutions." (p. 51) It is important to note that this is not the same as saying that after three centuries of science we have more problems than we had before. We may have more problems, but that is not the fault of science. Think of the problems humans have after two thousand years of Christianity. The logic or lack thereof is the same.

While writing about the Anthropic Principle in cosmology, Caiazza avers: "Understood scientifically, what becomes apparent is the inherent improbability of human life and life itself in its biological sense." (p. 80) This is contrary to the current scientific understanding that life may not only be widespread in the universe but may be an inevitable consequence of the nature of matter and energy.

On page 81 Caiazza states that "...a new sense of possibilities which include alternate universes, anthropic principles, indeterminacy, and a dynamic not a static universe, leads to the possibility that God exists in signified relation to the universe and not merely as an outside observer." Of course these new "possibilities" indicate nothing of the sort. They are all possible without any sort of God.

In (speciously) trying to account for the evolutionary adaptability of religion in Chapter 10, Caiazza fails to mention the cohesiveness brought to the tribe from a shared religious view or to mention the fact that religion helps to get young men to die in battle for the tribe, thereby furthering the survival prospects of tribal genes. This is probably the most important reason religion is universal among humans societies.

I should also point out that some of the chapters are irrelevant or dimly tangential to the war between the fishes. Especially out of place is the chapter entitled "The Agony of J. Robert Oppenheimer."

Finally I did not care for this, which in the tradition of the religions of the Middle East makes much too much of sex: "Sexual orgasm is the most intense single sensory experience a person can know and once removed from its social packaging of modesty, awe, morality, and family life, there is no ethical principle which prevents the pursuit of experience of sexual behavior at any time or of any variety." (sic) (p. 151)

Personally I am on side of the Darwinian angels since in the final analysis the only methods of acquiring knowledge that are not scientific are appeals to authority or to faith. But as Emily Dickinson wrote: "'Faith' is a fine invention/When Gentlemen can see/But Microscopes are prudent/In an Emergency."

Dowd, Michael *Thank God for Evolution! How the Marriage of Science and Religion Will Transform Your Life and Our World* (2007) *****
A most remarkable book

Whether Michael Dowd will succeed in reconciling the ways of the ancient religions to the facts of postmodern science, and in doing so, transform our lives by ending the dangerous contention existing between and among the various claims to "the way, the truth, and the life," remains to be seen. He is aiming for nothing less than the complete consilience of science and religion, a merging that, if successful, will be of inestimable value to humankind. I greatly admire the wisdom and intelligence and learning that Dowd brings to this very difficult task. I am amazed at his creativity and his temerity. His idea reminds me of something relatively simple, yet earth-shaking, something that might come from an Einstein or a Gandhi. I am not exaggerating.

The idea is this: we can accept as public truth and as "daytime" knowledge the facts about our world and ourselves as revealed through physics, cosmology, evolutionary psychology, cognitive science, neuroscience, geology, etc., while maintaining our faith in our religious heritage. We can still believe in Jesus Christ as our savior and be guided by the wisdom in the Bible while knowing that the earth really is four and a half billion years old and that, yes, we did indeed evolve from a long extinct ape-like creature.

It might be that Dowd is inventing a discipline. Call it Evolutionary Theology. Because we are educated we know that evolution is a fact; and because we believe in a God who cares and is intimately involved in this world, we therefore must see evolution as God's way of working in this world. But can the denotative words of the Bible be reconciled with such an understanding? Dowd's way around this conundrum is to understand that the Bible, inspired by God, was written in a way comprehendible to the people at the time, using words and images and ideas consistent with their world view. To write in the way of the modern world with the modern understanding would be unintelligible to those people and counterproductive.

This is a nice dodge (if I may) with some plausibility. I am satisfied with just saying that where the Bible is denotatively wrong, it is agreeable to interpret it symbolically. Dowd shies away from this direct approach because it would not help him with his consilience since evangelicals and others who believe in the literal truth of the Bible are sworn enemies of symbolic interpretations.

Dowd wants to celebrate evolution as our "cherished creation story." (p. 37) He sees facts as "God's native tongue." (p. 68) He makes a distinction between the "day language" of fact and the "night language" of meaning, between public revelation and private revelation, between reason and reverence (see especially p. 104). In this way differing utterances and experiences can be reconciled. I was especially enthralled because a friend of mine had the most intense dreams and visions in which she saw truths about the "other side" that she wanted so much for us all to accept. My way of accepting her views without compromising my own beliefs and experiences, was to refer to "public truths" and "private truths." No one can deny your experience. It is "true," but it is a private truth. Of course some people want more than that. They want their truth to be the public truth, and therein lies a problem of immense force: think of the differences between Christianity and Islam, between both of them and, say, Buddhism.

Dowd defines God as "the Ultimate Whole of Reality" (p. 77) and a wonderful definition it is! How tiny, how petty, how insignificant and sadly anthropomorphic seem the lesser gods! Dowd writes, "God cannot be limited to the world we humans can sense, measure, and comprehend: Ultimate Reality transcends and includes all that we can possibly know, experience, and even imagine." (p. 109) He goes on to reveal that the God he believes in is like the God of the Vedas, Ineffable and indescribable: "Any 'God' that can be believed in or not believed in is a trivialized notion of the divine." (p. 109)

Dowd calls the Big Bang of cosmology the "Great Radiance," and again what a way with words and ideas he has. He involves us all personally with the cosmic act of creation by reminding us that we are star dust, that we are the universe becoming conscious of itself. This identification with all of creation is a marvelous thing. Instead of narrowing identifying with only our group or nation or religion how much better it is to identify with the entire cosmos. There is great sense of freedom and wonder in doing so, and how petty seem these worldly conflicts when measured against the stars.

One of Dowd's most compelling and wondrous ideas is to recognize that the entire universe is evolving. He writes, quoting physicist Brian Swimme, "Earth, once molten rock, now sings opera." (p. 121) And we are an integral part of that evolution. Instead of being alone in a vast, uncaring, mindless universe, we are "a mode of being...an expression of the Universe. We didn't come into the world; we grew out from it, like a peach grows out of a peach tree." (pp. 120-121)

In short, what Michael Dowd has done in this remarkable book is to reconcile science with the tenets of the ancient religions, especially the Christianity he was born into. In a sense this a distinction between what he calls "flat-earth" Christianity and "evolutionary" Christianity. Throughout Dowd demonstrates a strikingly thorough understanding of evolutionary psychology, cognitive science and neuroscience, not to mention cosmology and even some physics. I say "strikingly" because it is so rare for someone formally trained in theology to have such a broad education. After this book achieves the kind of currency I expect it to achieve, perhaps the clergy will be respected (as they once were) as truly knowledgeable people.

Humes, Edward *Monkey Girl: Evolution, Education, Religion, and the Battle for America's Soul* (2007) *****
Fine account of a landmark case

The "Monkey Girl" in the title is one of the victims of the gross ignorance enacted by the Dover (Pennsylvania) Board of Education in 2005 as its majority members tried to slip creationism into the science classroom. She was a student who was made fun of because her parents accepted the empirical truth of biological evolution. It's an apt title for this very readable account of the Kitzmiller v. Dover "second Scopes trial" that ended in a victory for proponents of evolution. It's appropriate because ad hominem attacks, non sequiturs, and outright stupidity are characteristic of the style practiced by the parents of children who call others "monkeys."

Edward Humes made his reputation as a true crime writer. He has now branched out into other fields, and I am very glad he has. He brings rare reportorial skills and great energy to any subject or event he writes about. He writes clearly and without any phony humbug, and he does his research. And he is admirably fair.

However, it's hard to say just how "fair" this account may seem to those who believe in Intelligent Design or creationism or, for that matter, in the literal interpretation of the Bible. The plain fact is that ID is not science, and a literal interpretation of much of the Bible is blatantly false in a scientific sense. Furthermore, although it is difficult to say nice things about True Believers in this sorry context, amazingly enough Humes does just that. Even though the real instigator of this embarrassment to the Dover school district, onetime Board

President Bill Buckingham, revealed himself as someone who would (and did) lie under oath, who is grossly ignorant about matters of science and faith, who doesn't play fair and bullies people—Humes nonetheless describes him in a way that elicits sympathy for him.

One does not however feel sympathy for the Discovery Institute with its Intelligent Design Trojan horse "wedge strategy" with which it hopes to replace science in the classroom and ultimately, as Humes has it, win America's soul. Such people are playing for power. They want control. They want to replace the scientific method with appeals to authority for whom they hope to speak. They want to put God in charge so that they, as those who are in a position to speak for God, have the power and the authority. They are like Pat Robertson and his ilk who believe that any lie (and ID is a lie) is okay if it is done in the name of God.

Humes not only covers the Dover trial in depth but he gives the historical context in which it occurred as well as information about previous trials covering similar circumstances, including the famous Scopes trial and the precedent setting Edwards case from 1987 in which the Court ruled that creationism is religion and cannot be taught in the public schools. This is the case that got the Discovery Institute and others to come up with ID to further their anti-evolution agenda. Intelligent Design has been described as creationism in a tux. The conservative and Republican Judge John E. Jones III who presided over and ruled in the Dover case decided that ID was simply creationism (no tux), and that is basically why the Dover Board lost the case.

Humes makes not only the principals in the case come to life as he describes their actions, but he also makes crystal clear the legal and scientific arguments, which to the uninformed may be difficult to follow. This is a good book for people who are not experts who want to know what happened in Dover but also want to understand the issues involved.

Curiously enough Humes' epilogue contains a devastating mini-review of Ann Coulter's book "Godless." Humes waded through it (something I would not take the time to do) probably because one of the intellectual leaders of the ID movement, William Dembski, helped her with the science. Humes quotes a three-sentence passage from Coulter in which he identifies "five lies and one ludicrous error." (p. 346) In doing so Humes makes what I think is an important point that clarifies a lot of what this struggle is about. He writes:

"Perhaps the most outrageous lie contained in this three-sentence passage is Coulter's claim that liberals think evolution disproves God. In truth, the exact opposite is true: It is conservatives who think this way. Religious conservatives, not liberals, have tried to ban evolution from the public schools for decades because it contradicts their literal reading of the Bible." (p. 347)

Scientists in general and especially evolutionary biologists in particular know that evolution says exactly nothing about the existence or non-existence of God.

It is an irony of the times in which we live that those who most strenuously claim to be Christians are the ones who tell the biggest lies and tell them most often.

Hunter, Cornelius G. *Darwin's God: Evolution and the Problem of Evil* (2001) **

Specious arguments against evolution

Cornelius Hunter's thesis is that Darwinism and the theory of evolution in general rest on metaphysical "presuppositions" that are themselves unscientific, and that therefore Darwinism is really theology in disguise. To support his case Hunter repeatedly makes the argument that Darwin and various evolutionists are saying in effect, "God would never have made the world this way; therefore there was no divine creation."

Although some evolutionists, usually through carelessness of expression or contextual ambiguity, have made statements similar to that—Hunter quotes Stephen Jay Gould as writing something similar on page 48—when they have, they are mistaken, just as Hunter is mistaken in supposing that such an argument underpins Darwinism.

Evolution has a lot more going for it than a specious argument. Hunter is aware of this and in the course of the book tries to cast doubt on the overwhelming tide of evidence for evolution from the fossil record through molecular biology. Here's a typical example from page 38: "We have no idea how

48

the genetic code originated; therefore we can hardly appeal to its existence as evidence for evolution." But that doesn't follow. I may not be able to account for the origination of the rock that went through my window, nonetheless I can appeal to it as the proximate cause of the broken glass. And on pages 31-32 we find, "At the core of evolutionary theory is Darwin's law stating that in most instances it is the fittest that reproduce. But due to the complexities of nature and its life forms, we usually cannot measure fitness aside from counting offspring. Those organisms that leave more offspring are usually more fit, but we are not sure precisely why." Here Hunter reveals that he doesn't understand that evolutionary fitness is defined strictly in terms of reproductive success, period. It has nothing to do with "complexities of nature," and there is no more precise way to measure fitness.

Hunter also argues at length that Darwin was led in part to his theory of evolution through a desire to "reconcile the ways of God to man" (Milton) and especially to account for the existence of evil in this world. Again this is specious. Darwin may have been led, in part, to his theory of evolution because of his religious beliefs, but that has no bearing on the validity or effectiveness of the theory of evolution. Indeed, whatever Darwin's motivation was, it is irrelevant to the validity of evolution.

However the main fault of this book is simply a misrepresentation of just what it is that has made evolution the enormously persuasive theory that it is. Hunter writes on page 162, "But in fact the theory of evolution relies on the belief that God never would have created the world as we find it." But this is emphatically NOT what evolutionists are saying. It's not that God would never have created the world this way (or any other way, for that matter) but that the intricacies of the fossil and molecular record are better explained by evolution than by an appeal to a metaphysics. God might have divinely created everything in seven days and made it look like billions of years. That supposition can never be disproved. It is also the case that the moon could be made of green cheese and the experience of NASA and our scientific instruments are being fooled by the Green Cheese God. If I say that "God wouldn't work that way" (as Hunter accuses evolutionists of saying), I would indeed be committing myself to knowledge I can't possibly have, and if I say this is proof that the moon is made of rock and mineral, etc., I have made a simple logical error. But I would not say that, and neither do evolutionists say (if they are careful with their words) that evolution is proven because God would not work in such and such a way. What IS being said is that the report of our senses is better evidence than an arbitrary appeal to metaphysics, which is exactly the way science cannot work. The Green Cheese God may indeed exist and he may be fooling us to test our faith in him, but to paraphrase Damon Runyon, that ain't the way to bet.

Shanks, Niall *God, the Devil, and Darwin: A Critique of Intelligent Design Theory* (2004) *****
Demolishes the modern argument from design

Professor Shanks has done somebody a real service here in painstakingly demonstrating the utter intellectual poverty of so-called "intelligent design theory." Just who that person is I don't know. Perhaps it's a US congressman. Most people I know either haven't a clue about the subject, or are rationalists and are well aware that the intelligent design argument is scientifically vacuous and actually a religious power play, or they are religious true believers themselves and uncritically accept the notion that the universe was designed by a supernatural being whom they call God.

In other words, all the close and detailed analysis done by Shanks in this book—and trust me, he really addresses the question in the most thorough way—isn't about to persuade anybody one way or the other. Most people won't—and could not even if they tried—read it. It is entirely too finely meshed in technical detail about matters of no particular interest to them: cosmology, quantum mechanics, probability theory, biochemistry, thermodynamics, etc. Yet the book had to be written just for the record, one might say. All the pseudoscience served up by the creationists and the intelligent designers needed to be answered thoroughly, and Shanks has done that in a most impressive manner.

Shanks takes the intelligent designers seriously and presents their arguments, and then, piece by piece, refutes them. Frankly, I believe he gives them more attention than they deserve. After all, how seriously can one take a man (leading intelligent design theorist, William Dembski, for example) who writes: "My thesis is that all disciplines find their completion in Christ and cannot be properly understood apart from Christ"? (p. 157) I mean, isn't it enough to just quote such a person? He's a true believer and all his "arguments" are merely attempts to justify his belief in a supernatural being and supernatural causation. No amount of counter argument from logic or scientific experiment or from the multitudinous conclusions of the various sciences is going to sway him one iota.

But of course Shanks is not aiming his arguments at Dembski or his colleagues. Rather, like the good teacher he is, Shanks wants it spelled out for his students and for students everywhere just how absurd and wanting is the case for intelligent design. He is writing for those not yet entirely corrupted by religious propaganda and as yet innocent of the weight of the scientific evidence.

Why, one might ask, are the religious fundamentalists so intent on attacking Darwinism? Is it because they are uncomfortable with being closely related to apes, as were the Victorians? They probably are, but the real reason is that "Darwin's theory of evolution can be viewed as a sustained refutation of the argument from design..." (p. 24) Before evolution it was a mighty mystery as to how species arose, and any argument was as good as another, with the hoary argument from design being especially agreeable; and therefore pronouncements from the clergy held not only psychological, social and political sway over the masses, but intellectual sway as well. Darwinism changed all that, with the result that the Church lost an enormous amount of power and prestige—power and prestige that it has been desperately trying to regain ever since.

Noteworthy is the fine introduction by Richard Dawkins who has fought long and hard himself against the stupidities of the creationists and intelligent designers. Note well his sharp and decisive tone: "Intelligent Design 'theory' is pernicious nonsense which needs to be neutralized before irreparable damage is done to American education." (p. x)

That really is the bottom line. All that we have learned from science and rationalism is under attack from the forces of ignorance, mostly right-wing religious fundamentalists who would substitute their authoritarian mumble-jumble for reality in an attempt to seize the reigns of political power and usher in a return to the Dark Ages with themselves at the throne. Professor Shanks is to be commended for his efforts to prevent such a catastrophe, as unlikely as such a catastrophe might be.

Williams, Robyn *Unintelligent Design: Why God Isn't as Smart as She Thinks She Is* (2006) *****
Part rant, part memoir, part science and philosophy

Robyn Williams is a man of science, a broadcast and print journalist, and an academic who has joined the chorus of voices that are fed up with the attempt by creationist and Intelligent Designers to hijack our societies. Williams presently writes from Australia but has lived in Europe while growing up in the UK. He focuses on American culture and American creationists as well as on those from Down Under and in Europe. He is witty, glib, chatty, and quick with a satirical rapier worthy of Voltaire.

He sees the "unintelligent design" movement fired by "proud ignorance" and a lust to power. I couldn't agree more but I'd ratchet it up to "arrogant ignorance." Williams means "proud" as in the sort that cometh before a fall. I have little doubt that the creationists and their turkeys in tuxedos, the IDers, will go the way of the dodos eventually as our populations become educated and no longer easily swayed by the charlatans of religious mumbo jumbo.

The book is jeans and t-shirt causal. There are no footnotes or endnotes or a bibliography or an index. He quotes whole passages from people like Richard Dawkins and Jared Diamond but doesn't say exactly where he got the words. He mixes memoir with secular sermon (no soda water), history with incident, and passion with the jocular. He ridicules the notion of an anthropomorphic God, asking if God (in whose image we are said to be made) ever gets a bad back. The very idea of intelligent—intelligent!—design is made absurd. He writes, "Halitosis, farting, vaginal discharge, reflux, snoring, rheumatism, warts, smelly armpits, varicose veins, menopause, brewer's droop...these are not the marks of a designer at the top of his game." (p.71)

Williams spends some ink debunking ID claims to life from complexities so great that they had to have a designer (the main ID delusion). Here he brings Richard Dawkins into the fray in a most delightful way:

Dawkins in his book *The Blind Watchmaker* is busy rebuking the Bishop of Birmingham, Hugh Montefiore, author of *The Probability of God*. Dawkins quotes Montefiore: "As for camouflage, this is not always easily explicable on neo-

Darwinian premises. If polar bears are dominant in the Arctic, then there would seem to have been no need for them to evolve a white-coloured form of camouflage."

Dawkins gives this mock translation of Montefiore's paragraph: "I personally, off the top of my head sitting in my study, never having visited the Arctic, never having seen a polar bear in the wild, and having been educated in classical literature and theology, have not so far managed to think of a reason why polar bears might benefit from being white."

Williams guides: "Predators need to surprise their prey." He adds, "Bishops, and lay folk too, may like to take the trouble...to see what science has come up with." (p. 57-58) If they do so they might not fall into what Dawkins has called "the Argument from Personal Incredulity," or what Williams has generalized as what I'll dub as "the Argument from Ignorance."

Much of the book is like this, and some of it is very funny indeed and works well as a scathing revelation of the stupidities of the Intelligent Designers. But there is also some very personal writing in the book that surprised me. Williams writes about being a child with a heavy-handed communist father and being in a secondary school in London in the 50s. In short this is a book that engages as much as informs or guides.

Finally I want to address this idea that Williams brings up on page 76. "...[C]onflict is creative and...isolated societies decline. As happened in Tasmania before the Europeans, the technology becomes more primitive and the people languish without invasion, rape and pillage to renew the innovative stock." This is connected with the (specious) argument that the seemingly evil God who is allowing all the carnage is just doing it for our own good. What I wonder is does humankind have the ability to survive the current domestication of some of our populations (think couch potatoes in America). When the war system ends (as I hope and trust it will—eventually) what is to become of the Eloi?

I don't know the answer to that question, but perhaps we will acquire the wisdom to redesign ourselves in a way that allows for greater human happiness over greater periods of time.

To close here is a nice Williams rant: "...[T]he human brain...resembles the creation of the devil rather than of a God. That it is capable of good is beside the point. ID is like a computer program with a built-in virus. ID is a baby born with syphilis. ID is an insult to the intelligence. ID is an insult to God." (p. 76)

Amen, brother.

Wilson, Edward O. *The Creation* (2006) ***
Admirable, but addresses the wrong people

This is an interesting book, as all of Wilson's books are interesting. I like the story about the fire ants that plagued the islands in the Caribbean some centuries ago and the paleoforensics employed to figure out what happened; in fact I like all of Wilson's stories about ants. Ants are fascinating, and ants will be here long after we are gone.

I also like the idea of trying to preserve as much biodiversity as possible. Save the rain forests, by all means. Save the tigers and the gorillas, the elephants and the snow leopard.

But it's not going to happen. Wilson is in prayerful mode. You can tell that by the very fact that he addresses this plea to a protestant clergyman as an author's conceit (partly in remembrance of his Baptist childhood). His tone, try as he might, will be taken by some as condescending, which it unavoidably is. Wilson is sensitive to the criticism he has gotten over the years, especially from those who think his sociobiology is a blueprint for a return to eugenics. So he is overly polite, overly indulgent with all the references to the Bible, to what he and the clergyman have in common. He is bending over backwards.

But it will never work. Biodiversity means little to the average clergyman. Saving the planet and its resources may mean a little more since even though evangelical Christians are certain that the rapture is coming, they are uncertain as to when. It could be tomorrow; it could be a few years away. So let's not dirty up the waters too much, let's not kill all the honeybees, let's scrub the coal smokestacks; in short let's not allow an environmental holocaust, at least not yet.

This is a short book, the kind of book that eminent persons are allowed to write and see published near the end of their lives, the kind of book that speaks of public service, that adds some further meaning to the author's existence. In a way this is similar to *Consilience* (1998) in which Wilson called for a meeting of the minds between the hard sciences and the soft ones, between those in the humanities and those in the labs, for reductionism to embrace poetry or vice-versa. The same idea applies here: people of faith should join people of science and work together to preserve the biosphere.

However, instead of addressing clergymen I think it would have been better and more natural for Wilson to address heads of state and chief executive officers of giant corporations since they are the ones most directly responsible for the ecological disaster staring us in the face, and they are the ones who can do something about it. Wilson writes, "Humanity doesn't need a moon base or a manned trip to Mars. We need an expedition to planet Earth...." Clearly such a statement would be better aimed at the Congress of the United States than at its clergymen. Wilson indicates in this book that he is not much for genetic engineering or for saving humanity by moving into outer space. He writes, "...human biology and emotions will stay the same far into the future, because our immensely complicated cerebral cortex can tolerate little tinkering..." (p. 28)

I also think it is unfortunate and even obsequious that Wilson calls what he wants saved "The Creation" when it is obvious that he does not consider life on earth a creation at all and in fact states directly that life on earth evolved from nonliving matter and energy. This sop to the creationists and Intelligent Designers is somewhat offset by his argument against Intelligent Design in the last chapter.

This book could also have been addressed to young students and teachers of biology, which in fact is what he effectively does in chapters 12-15 which are titled, "The Fundamental Laws of Biology," "Exploration of a Little-Known Planet," "How to Learn Biology and How to Teach It," and "How to Raise a Naturalist."

Putting aside the artificial spin that Wilson employs, this book is really about "three problems that affect everyone: the decline of the living environment, the inadequacy of scientific education, and the moral confusions caused by the exponential growth of biology." (p. 165) Wilson addresses the Christian clergy because "In order to solve these problems...it will be necessary to find common ground on which the powerful forces of religion and science can be joined." (p. 165)

More in keeping with the Wilson I know and greatly admire, typified in his book *On Human Nature* (1978) which won the Pulitzer Prize, is this from page 28: "There are still some thinkers around the world...who wish to base moral law on the sacred scripture of Iron Age desert kingdoms while using technology to conduct tribal wars—of course with the presumed blessing of their respective tribal gods."

I think the average clergyman, here and elsewhere, is still in the thrall of his tribal god, and not likely to listen to Professor Wilson, regardless of how politely and diplomatically he presents his case. Too bad.

Bottom line: clearly not one of Wilson's better books. I might even say that he has lost his intellectual compass. I hope he finds it.

Chapter Three

Evolution Writ Small

What I mean by "writ small" is that the focus here is on the evolution of tiny entities like prions, viruses and bacteria, and on the microscopic elements of cells such as DNA, genes, proteins, and mitochondria. Also examined are ideas about the role that epigenetics and biochemistry play.

Perhaps one of the most important ideas is how the cellular environment and random "developmental noise" determine how the DNA code is read and how this determines what the phenotype will be like. Clearly more knowledge about how DNA and the developmental environment work together to grow organisms is crucial to our understanding. Another important idea is how DNA analysis supports the truth of biological evolution.

Avise, John C. The *Genetic Gods: Evolution and Belief in Human Affairs* (1998)
*****
Engaging exploration of genetics and attendant ethical issues

John Avise's engaging book is both an overview and an introduction to recent genetic research as well as an assessment of the social, ethical and religious ramifications stemming from our manipulation of the genetic code. The terminology is a little formidable in spots, and there is perhaps more genetics being explained here than most general readers would want, but these are minor obstacles when one considers the reward: listening to an expert talk about what's happening in genetics today while considering the implications. I was very impressed with Avise's level-headed and balanced assessment of the controversies. This is a sophisticated book, deeply considered and carefully expressed.

The author is the distinguished Professor of Genetics at the University of Georgia and an evolutionary biologist who really knows his stuff. "The genetic gods" in the title is a metaphor of course—he even refers jokingly to "protein angels" on page 208—his point being that we are to some very real extent at the whims of our genes, just as the Greeks once thought they were at the whims of the gods on Mount Olympus. However don't imagine that Avise is presenting a genetic-centered reductionist approach in this book. He wants to emphasize that the genes are subtly, and sometimes not so subtly, influencing our lives, but that is far from the whole story. Avise suggests that the proper way to look at the culture verses genes debate is to think of culture as an "epigenetic phenomenon...itself a product of biological evolution," and that "genes and culture coevolve" (pp. 158-159). The environment shapes us, but we in turn condition the environment. As Avise expresses it, "the individual's mind to a considerable extent creates itself through the environments it conditions." (p. 159). In our attempt to understand how the mind works and to account for human behavior, Avise's states that a "myopically reductionist approach that neglects multiple levels of biological, personal, and social causation" is inadequate, as are "uncritical holistic approaches alone" (p. 165).

This just makes sense and it also makes moot the sometimes heated "culture verses genetics" debate, which is similar to the old "nature verses nurture" false dichotomy. Quite simply, how can we separate the effects and influence of the environment and culture from that of the genes, and vice versa?

Professor Avise does not shy away from a position on whether we should intervene genetically when something is amiss. While some people believe that "developing embryos are governed by intelligent and caring supernatural forces," Avise insists that "they are governed by natural gene-environment interactions that unfortunately can" go horribly awry as in the Lesch-Nyhan syndrome or in Down's syndrome (p. 199). In such cases, he asks, "Should we then take the reins?" His careful answer on the next page is yes, but with the understanding that "the interests of the individuals most closely involved...should take legal precedence over those of more distant parties."

Avise believes that ethical questions about genetic engineering should be considered by all members of society, not just scientists or theologians or lawyers (p. 202). He believes in a case-by-case appraisal (p. 201). "The only <wrong> approach," he avers, "is that in which the moral authority of a god is asserted." He wryly observes that since there is such a diversity of opinion, "any supernatural deity either has been strangely silent on such issues or else has conveyed vastly different messages to different listeners."

This book requires an effort on the part of the reader, there is no doubt about that. This is no breezy *Time* magazine treatment. But I think what we can learn from Professor Avise on a topic of such overriding human interest is well worth the effort.

Boaz, Noel *Evolving Health: The Origins of Illness and How the Modern World Is Making Us Sick* (2002) ****
Excellent introduction to the ideas of evolutionary medicine

This works as a general introduction to the nascent field of evolutionary medicine. Note well the word "health" in the title. One of the central ideas in evolutionary medicine is preserving health, and in general looking at medicine from the point of view of the healthy instead of from an overweening concentration on the sick. An ounce of prevention in evolutionary medicine is worth a whole ton of cure.

Another important idea is to look, in so far as possible, to our adaptations as evolutionary beings to see what we might be doing wrong today. For example, grasses with plump seeds of carbohydrates were in short supply before the advent of agriculture about 10,000 years ago. There were wheats and ryes, wild oats and such, but their seeds were relatively small and required a lot of labor to harvest. Consequently, our ancestors on the savannahs and in the woodlands ate grain carbohydrates in small amounts. Now, of course, grains—especially rice, wheat and corn—are the staple foods everywhere in the world and we eat massive amounts of them.

Is this a problem? As Professor Boaz points out, evolutionary medicine suggests that it is. We are "carbohydrate intolerant" (Boaz uses the term "glucotoxicity," page 133) and cannot shut down our appetite for all the carbohydrates so tantalizingly available to us. They are especially enthralling when served up with salt and fats.

In the prehistory there were no supermarkets open 24-hours a day. Instead there were freezing winters and droughts that might last for months or more, sure to visit almost every human eventually. So when there was a bountifulness in the land we ate our fill and then some. And those of us who had the ability to put on fat could live out the times of famine better than any prehistoric runway model. And so our chubby guy- or chubby gal-genes were favored. Boaz calls this the "thrifty genotype."

However that virtue has become a fault. What to do? Boaz recommends exercise, for one thing. In the pre-history our ancestors managed to walk all the way around the world. They had no cars or easy chairs. That we can solve our fat problem by looking at the way our ancestors lived and emulate them, is the somewhat bitter pill of this book. And, by the way, this "medicine" (hard to take, as we all know) also works against heart attacks, gout and other modern diseases.

Boaz has gone to some considerable trouble to associate various "diseases" with 17 evolutionary levels of human structure and function. (There's a table on pages 19-25.) These levels are like the idea that "ontogeny recapitulates phylogeny" in that some of the levels are similar to those stages in the embryo's development from single cell through bony fish and amphibian to mammal, all the way to us. What Boaz is adding here is the idea that certain diseases are associated with each level of development. For example, emphysema is associated with the amphibian level of adaptation while viral infections go all the way back to when our ancestors were just single cells.

This scheme is useful in helping us to understand disease. It is even helpful in treatment. But Boaz's formulation is no magic pill or cure-all. For the chronic diseases that plague those of us in the developed world there is no easy cure. Boaz recognizes a "discordance" between our evolutionary selves and the modern environment that is leading to these diseases. He uses a concept he calls "adaptive normality" that can guide us away from the discordance.

This is a very readable book requiring no prior expertise. It is obvious that Boaz wanted to reach the educated lay person with his ideas. For those of you new to the idea of evolutionary medicine, this will be an exciting book. Boaz does an excellent job of teaching us is how to think from an evolutionary perspective, which is something we all need to do.

Carroll, Sean B. *The Making of the Fittest: DNA and the Ultimate Forensic Record of Evolution* (2006) ****
How DNA supports the fact of Darwinian evolution

Most of this book is a primer on how the study of DNA code from various species sheds light on the evolutionary process. The text is as clear as such a text can be considering how abstract the DNA sequences are to even the well-educated reader. There are numerous charts and tables, drawings, black and white photos (and some color plates) and such in this timely, handsome and well-presented book to guide the reader. I only wish that I could have grasped the details in a more concrete manner.

DNA codes for proteins, of which there are vast numbers. These proteins are formed from amino acids of which life uses twenty, and in turn these amino acids are called up by the sequence of letters in the code. Presumably (Carroll does not make this clear) as the zygotic cell divides, working its way to the composition of the complete organism, the DNA code is read in sequence like a tape fed into a bar code. First this protein and then that protein and then still another is made and somehow strung together in an exacting order so that, voila! a massively complex organism is constructed. What is not in this book is an explanation of how these proteins know where to go and when. Presumably that knowledge is part of the very sequence of the code, or perhaps it is implicit in the positions in space of the proteins relative to one another. In others words, the DNA code is only the most obvious and "visible" part of the microscopic reproductive process.

If you are like me and are looking for the same sort of explanation, this book will be of limited value. Prof. Carroll's purpose is not to make transparent the reproductive process at the chemical level. His purpose—and a laudable one it is—is to show how DNA analysis is yet another piece of evidence pointing to the truth of biological evolution. That is why he uses the word "forensic" in the subtitle.

One of the most powerful uses of DNA is in reconstructing the so-called "tree" (or "bush") of life. Carroll shows how it is now clear beyond almost any doubt that our closest relatives are chimpanzees and bonobos followed by the other great apes and then monkeys. He shows how DNA analysis can also (and by the same logic) be used to show the relative age of species. Interesting is the discovery of how exactly similar are some sections of code in diverse species, indicating that such code is very ancient. In fact Carroll points to "immortal" sequences of code that have resisted all attempts at corruption or mutation. He explains that such code is so nearly indispensible to living forms that natural selection is, and has always been, active in keeping it intact.

In this regard (and moving to the latter chapters of the book) we find a particularly delightful refutation of one of the notions of "Intelligent Design." Carroll quotes perhaps the best known of the intelligent designers, Dr. Michael Behe, as writing:

"Suppose that nearly four billion years ago the designer made the first cell, already containing all of the incredibly complex biochemical systems discussed here and many others. (One can postulate that the design for systems that were to be used later, such as blood clotting, were present but not 'turned on.' In present-day organisms plenty of genes are turned off for a while, sometimes for generations, to be turned on at a later time)." (p. 244)

How brilliant this sounds! However Carroll writes:

"This is utter nonsense that disregards fundamentals of genetics. Dr. Ken Miller of Brown University has described this scenario as 'an absolutely hopeless genetic fantasy of pre-formed genes waiting for the organisms that might need them to gradually appear.' As we saw in chapter 5, the rule of DNA code is use it or lose it. The constant bombardment of mutation will erode the text of genes that are not used, as it has in icefish, yeast, humans, and virtually every other species. There is no mechanism for genes to be preserved while awaiting the need for them to arise." (p. 244)

Indeed, if Behe were correct, there would be in virtually all species "silent pre-formed genes" waiting to be called upon. There aren't.

In the chapter entitled "Seeing and Believing" Carroll recalls Louis Pasteur's struggle to demonstrate to non-seeing and non-believing doctors that child-bed fever was caused by their dirty hands. He reprises the horrific and bizarre story of the Soviet head of (political) biology Trofim Denisovich Lysenko who denied genetics, and how Stalin's support of him led to massive failures in agriculture and subsequent starvation. He further recalls how Mao Zedong, using the same unscientific ideas, sponsored massive starvation in China due to crop failures.

What Carroll is getting at is that political corruption of science can be very dangerous. In the United States today under the power of the Bush administration, faith-based (and corporate-sponsored) science is denying global warming and other deleterious effects of rampant pollution. This sort of science denial is likely to lead to human suffering and death, just as did the communist denial of genetics.

Cornish-Bowden, Athel *The Pursuit of Perfection: Aspects of Biochemical Evolution* (2004) ****
Difficult but rewarding

One imagines that in the great expanse of time from the formation of the earth until before the first fossils appeared (something like a couple of billion years) a veritable riot of biochemistry took place as proto-cellular life searched for the most efficient and economic chemistry to run its replicators. And as these replicators gathered other molecules to form amino acids and proteins, helped along by molecules serving as maidservant-like enzymes, their biochemical reactions became more and more efficient until self-sustaining cells featuring "organizational invariance" were well established.

Cornish-Bowden calls this process "the pursuit of perfection." He asks the question, are metabolic pathways and the enzymes employed in living systems optimized, or is it the case that the evolutionary mechanism hit here and there on something that worked, and stayed with it? In other words, are the ways that living cells use chemistry to run their systems Panglossian? Is the chemical world inside the cell, in general, the best of all possible ways it could be?

Cornish-Bowden, who is director of research at the *Centre National de la Recherche Scientifique* in Marseilles, France, argues that such is indeed the case. (See his direct statement on page 14.) His technique, relying to some extent upon the work of Spanish biochemist Enrique Melendez-Hevia (to whom this book is dedicated) is to postulate mathematically what an optimal metabolism might be and then to compare that to how the cell actually operates. Cornish-Bowden points out that optimal metabolism in the cell is not necessarily the most efficient in the lab since the reactions within the cell are constrained by a complex environment that includes hundreds of other reactions taking place at the same time.

Part of Cornish-Bowden's purpose is to demonstrate that, although enormously complex, biochemistry is not so complex as to require design by a supernatural being. Indeed, he argues that biochemistry is the direct result of how matter and energy work, and that some calculations of how "unlikely" it is that amino acid sequences used by living systems could arise by chance, are exaggerated. I think it would have been good if Cornish-Bowden had also emphasized that chance really has nothing to do with it, since many creationist and intelligent designer types still think evolution is postulated to proceed by chance. It does not. The majority view among scientists that I have read (and Cornish-Bowden is certainly among them) is that the biochemistry of life, however complex, is a direct result of the nature of matter and energy as they react in certain environments. The exact manifestations of the evolutionary process are contingent (the word that Stephen Jay Gould preferred) on the unraveling through time and space.

Another of Cornish-Bowden's purposes here is to present his ideas to a more general readership than biochemistry usually receives. He takes care to make the text as readable and accessible to the general reader as possible. In this, however, I think he is up against some nearly insurmountable difficulties. The problem is that, however carefully the author presents complex arguments involving technical matters using specialized terminology, that author, however talented, will not reach a general readership because the reader will not have the time or the inclination to *study* the arguments. And, alas, many of Cornish-Bowden's arguments are highly technical and require study to fully appreciate.

Furthermore it is almost impossible for a working scientist to write for a general readership within his area of expertise without being aware that other scientists in his field are looking over his shoulder. Consequently, the scientist is careful to make the many qualifications necessary and to cite the exceptions—in other words, to be rigorous. And rigor in science requires dotting i's and crossing t's in a manner that will often appear as Greek to the non-specialist.

For myself (your typical, limited, non-specialist) I took his detailed explanations at face value. I turned the pages when the argument got too specialized, and ultimately I concerned myself with his conclusions.

Here are some interesting ideas that I gleaned along the way, ideas that make this book definitely worthwhile:

Most important perhaps is the idea that advances in understanding biochemical evolution are leading the way toward a more powerful understanding of how evolution works. For example, it is the study of DNA, augmenting the sketchy fossil record, that has fixed modern human origins in Africa between fifty and a hundred thousand years ago.

Another interesting idea that Cornish-Bowden emphasizes is that "the sort of perfect correlation [between prediction and result] that would delight a physicist should arouse suspicion in a biologist." (p. 146) In order words, biology is a bit messy and inexact in comparison to physics, and should a one-to-one correlation arise, it might be the case that the correlation is a tautology or a mathematical artifact.

I also liked his very intriguing description of cancer as a "parasitic species that lives out its entire evolutionary history from speciation to extinction during a fraction of the lifespan of its host." (p. 150)

Important too is the distinction that Cornish-Bowden insists upon between "complicated" and "complex." Noting the property (from complexity theory) of "emergence," he explains that a "system is complicated if it contains many components...," but to be <complex> "a system needs...in some sense to be more than the sum of its parts." (p. 130)

Finally, I liked the way he explained how bacteria find food, a kind of movement he calls "a biased random walk." In essence the bacteria moves onward if things are getting better, but when things are getting worse "it stops quickly and tumbles," thereby randomly changing direction. (p. 50)

Harold, Franklin M. *The Way of the Cell: Molecules, Organisms and the Order of Life* (2001) *****
Difficult, profound; worth the reader's best efforts

Time and again in this dense, intensely scientific exposition on cellular life, Professor Harold expresses his dissatisfaction with what he calls the "genocentric" view of life. Instead he would like to see a "focus on the cellular templet rather than the molecular gene." He believes this would represent "a significant divergence from the genocentric conception of life that now dominates the scientific literature and even more so, the popular press." (p. 100) Harold makes a strong case for his point of view; indeed, it is this book more than any other that has made me see the overriding influence the immediate molecular environment has on reproduction and growth.

The genome has its "recipe," its code of instructions, but what Harold is at pains to tell us is that without the four-dimensional cellular environment in which the gene's "instructions" are carried out in a step-by-step process, there would be no growth or reproduction.

What this means is that the shape and temperature, the position and abundance of the surrounding cellular elements themselves shape the genetic expression as much as or even more than the genome. All life comes from life. All cells come from cells. There is no acting out of the genetic code outside a cellular environment.

And so we see Harold's frustration and that of other molecular biologists at all the hoopla that has accompanied the sequencing of the genome when it is clear that reading the code is just a very small step toward understanding how the cell reproduces itself and grows. What we need to understand is the intricate environment of the cell and how it interacts with the code leading to the epigenetic assembly of the cell and ultimately of the organism. The complications inherent in such an enterprise are truly mind-boggling in the extreme. Analysis of the four-dimensional factors would overwhelm the fastest computers in existence—all of them at the same time—if somehow we could figure out how to employ them to aid our analysis.

These facts explain why scientists like Harold are insistent upon a holistic approach to biology and why they again and again warn about the limitations of a reductionist approach. Life is just too complicated to be understood by breaking it down into pieces and attempting to put it back together, or to reverse engineer it.

On page 213 there is an interesting comparison of E. O. Wilson's view that there is "progress" in evolution and Stephen J. Gould's emphatic view that there is not. Harold seems to be implying that because organisms have become more complex that there is indeed at least "direction" in evolution. I would go further than this and observe that the rise of complex culture-bearing organisms like humans, who may be able to protect their home planet from a death-dealing meteor, implies if not "progress" in evolution, something equally agreeable. However, I would not say that our rise was inevitable. Indeed, along with Gould I would call it a contingency.

Much of the book, especially chapters three through eight, is a technical exploration of the microbial world of the cell using concepts and terminology not readily accessible to the lay reader. Harold is aware of this, at least for Chapter 4, "Molecular Logic," where he writes on page 35, "...students of biochemistry will find little in [the chapter]...that is new to them, but for the layman it may be like sipping water from a firehose." (!) Professor Harold provides a glossary, but one suspects one is out of one's depth when the words searched for are not in the glossary, but can be found in an ordinary dictionary!

Nonetheless the broad outlines of Harold's message can be discerned without appreciating fully the intricacies of cell metabolism and development. The introductory chapters, "Schrödinger's Riddle" and "The Quality of Life" explore the question that physicist Erwin Schrödinger famously asked in his much admired little book, *What Is Life?* (1944), a book that very much impressed the young Franklin Harold. In the closing chapters, beginning with Chapter 9 "By Descent with Modification," and especially the engaging Chapter 10 "So What is Life?", Harold looks more generally at evolution. He touches on the new science of complexity and how it relates to biology, and on the thermodynamics of ecosystems and how that affects natural selection. His treatment of some of the controversies in evolutionary theory is both illuminating and balanced, so much so that one would like to quote whole passages. This is obviously a subject Professor Harold has thought long and hard about for many years. Here are some examples of his thought:

"...[F]orm is not directly or rigidly determined by the genotype: the genes define a range within which the phenotype falls, but forms arise epigenetically as the result of developmental processes." (p. 209)

"Organisms are historical creatures, the products of evolution; we should not expect to deduce all their properties from universal laws." (p. 218)

"What we lack is an understanding of the principles that ultimately make living organisms living, and in their absence we cannot hope to integrate the phenomenon of life into the familiar framework of physical law. I am not here to advocate a veiled vitalism, nor to sneak in a creator by the back door. But...until we have forged rational links between the several domains of science, our understanding of life will remain incomplete and even superficial." (p. 218)

"...[W]hile a machine implies a machine maker, an organism is a self-organizing entity." (p. 220)

"Organisms process matter and energy as well as information; each represents a dynamic node in a whirlpool of several currents, and self-reproduction is a property of the collective, not of genes.... DNA is a peculiar sort of software, that can only be correctly interpreted by its own unique hardware.... [S]ending aliens the genome of a cat is no substitute for sending the cat itself—complete with mice." (p. 221)

Lewontin, Richard. *The Triple Helix: Gene, Organism, and Environment* (2000)
****

Why the genome project may disappoint

This little book contains three lectures given by Lewontin at the Lezioni Italiani in Milan a few years ago. It is technical and aimed at an educated readership. Since there is not enough space here to discuss the entire book, I will concentrate on a brief discussion of the first lecture, "Gene and Organism."

In this lecture Professor Lewontin outlines the role that genes, environment and chance ("random noise") play in the development of an organism. As he phrases it on page 20: "the organism is not specified by its genes, but is a unique outcome of an ontogenetic process that is contingent on the sequence of environments in which it occurs." This means that you could take the same genetic code and have it unfurl in Hyde Park and get an organism different from one you would get having it unfurl on, say, the Boston commons. Lewontin shows how cuttings from the same plant cultured at different altitudes developed differentially, and in a manner that could not be predicted. The reason they could not be predicted is that there is a significant amount of random variation ("developmental noise") that occurs as the plant grows. Lewontin gives the further example of a multiplying bacterium on page 37. The bacterium divides in 63 minutes. In another 63 minutes the daughter cells should divide again, giving four bacteria, but actually there is some random variation in how long it takes them to divide, so that one daughter divides in say 55 minutes, the other in an hour and five minutes. And this continues so that the bacteria culture does not increase in pulses, but continuously in random increments. This difference in timing in multi-cellar organisms may result in morphological differences since a catalytic enzyme may arrive too late to, say, grow a side bristle on a fruit fly (an example that Lewontin gives). Lewontin applies this understanding to the development of our brains on page 38. First there are random connections set. "Those connections that are reinforced from external inputs during neural development are stabilized, while the others decay and disappear." This process, Lewontin advises us, can lead to differences in cognitive function that are neither strictly genetic nor strictly environmental. They are influenced by random (unpredictable) factors.

This understanding is the reason that Lewontin is less than thrilled with the Human Genome Project. He believes, as he makes clear in another book, *It Ain't Necessary So: The Dream of the Human Genome and Other Illusions* (2000), that we will be disappointed by what can be accomplished simply from sequencing the genetic code, his point being that even though we know the code, the environmental and random factors cannot be known in any precise or predictive sense. It is true that the genome for a chimp will always code for a chimp and never for a rabbit, but whether that chimp is good at math or has unusually aggressive tendencies is something we cannot know from an understanding of the genetic code alone. Chance and environmental factors in development can result in a passive chimp even though its parents are aggressive.

Applying this idea to evolution in general, we can see that individual variation is not strictly a result of environmental differences but also of chance differences. Consequently, what we are is not shaped strictly by adaptive pressure (natural selection) but is to some extent the result of purely random processes. At one time in my life I studied chance and random events, and one of the most important things I learned is that the term "random" is not clearly defined, except in the sense that something that is random is unpredictable, which is a "you can't prove a negative" sort of definition. I also learned that there is considerable doubt as to whether a truly randomizing device actually exists. All real world devices, such as roulette wheels and computer random number algorithms can be shown to have some tiny bias, or to break down at the extremes. (Don't trust the random number generator on your computer when you are generating a very large number of trials: it will begin to repeat, and your Monte Carlo simulation will be flawed.) So what Lewontin calls "random events" are actually events that we simply do not know enough about to describe accurately. It may be that with greater ability we will eventually be able to describe or control these events. However, it may also be that at some level such events are the direct result of the probabilistic nature of a quantum event, and therefore in principle unpredictable. I suspect that Lewontin believes something like this.

In the second lecture Lewontin makes the point that to a significant degree organisms create their environment, and it is wrong to think of a place (such as the surface of the moon) without organisms as an environment. His dictum is "...[T]here are no environments without organisms" (p. 67). In the third lecture Lewontin discusses some of the problems associated with genetic causation and its analysis. There is a fourth chapter in which Lewontin attempts to provide some direction for future studies in biology.

I did not understand his assertion on page 81 that "Only a quasi-religious commitment to the belief that everything in the world has a purpose would lead us to provide a functional explanation for fingerprint ridges or eyebrows or the patches of hair on men's chests." The hair, I imagine is the result of sexual selection, but surely the fingerprint ridges allow us a better grip, and our eyebrows shade the sunlight as well as providing some small cushioning for our eye sockets.

Moalem, Sharon with **Jonathan Prince** *Survival of the Sickest: A Medical Maverick Discovers Why We Need Disease* (2007) ****
Understanding genetic disease from an evolutionary point of view

We really don't "need" disease. This is a bit misleading. It just so happens that some genetic disorders, such as sickle-cell anemia, favism, diabetes, hemochromatosis, the tendency to obesity, etc., confer on the afflicted compensatory advantages. Thus a predilection for getting fat is adaptive if a drought or a long winter beckons, or a person with a genetic tendency toward sickle-cell anemia is less likely to get malaria, and so on. Note that it is only diseases caused by genetic mutations that Dr. Moalem is talking about. (It might be true that we "need" disease in the sense that prey animals need predators—but that is another matter.)

One of the techniques our bodies use when fighting infection is to reduce the amount of iron available to the invaders. Bacteria need iron to reproduce. If there is a lot of it available their numbers can grow quickly. Without iron they can't reproduce at all. Iron is a limiting factor for many kinds of life. Vast stretches of ocean support little in the way of life because the microorganisms that begin the food chain can't grow where there is so little iron. As Dr. Moalem reports in this wide-ranging and eyebrow-lifting book, sprinkle some iron onto those patches of ocean and they will quickly turn green with microorganisms.

So it is a bit of an irony that people who have hemochromatosis, a genetic disorder that causes them to retain large amounts of iron in their bodies, are able to survival infections like the plague. This is because they starve the invading microbes through "iron locking." They have a lot of iron in their bodies, but they keep it away from the bacteria. Other people who have low levels of iron in their bodies are able to withstand bacterial attacks because they also keep what little iron they have away from the germs. In fact, one of the body's initial responses to microbial invasion is to limit the amount of free iron in the system.

Genetic coding for levels of iron in the body is an example of evolutionary adaptation, part of the ongoing arms race between us and the microbes that live in and on our bodies. This is just one of several interesting and new ideas coming from the growing science of evolutionary medicine that I found in *Survival of the Sickest*. Incidentally, one way to manage hemochromatosis is through donating blood on a regular basis, which explains in part why physicians of old were sometimes successful when they bled their patients.

This got me to thinking about "only women bleed" which led me to think about hemorrhoids (which prove that it isn't only women who bleed). Perhaps bleeding instead of retaining blood, which seems like the more natural thing for our bodies to do, has adaptive value in some people in some environments.

Another interesting idea is this from page 58: "ACHOO syndrome—its full name is autosomal dominant compelling heliopthalmic outburst syndrome." It is a "disorder that causes uncontrolled sneezing when someone is exposed to bright light, usually sunlight, after being in the dark." Dr. Moalem suggests that "way back when our ancestors spent more time in caves, this reflex helped them to clear out any molds or microbes that might have lodged in their noses or upper respiratory tract." Now this may sound a bit farfetched, but I have suffered from low grade allergies all my life, and used to have asthmatic attacks. I came to believe that the buildup in my lungs and the sneezing were signals to me to move on! Of course now I clean and vacuum like a germaphobe, but the idea is the same. My symptoms were adaptive. They more or less forced me to reduce the level of potential irritants and microbes in my environment.

But there is more. I noticed long ago that sometimes the sun in the morning would cause me to sneeze. I never figured out why until I read the above from Dr. Moalem. I am just the kind of person who would need to sneeze those molds out.

Later on in the book Moalem returns to an evolutionary idea that has been kicking around for decades. Beginning with the work of Elaine Morgan from the 1970s the public became aware of the notion that we humans had an aquatic past. She got the idea from marine biologist Alister Hardy. Through such books as *The Descent of Woman* (1972) and *The Aquatic Ape: A Theory of Human Evolution* (1982) Morgan argued that some of our unusual adaptations came about because we had an aquatic past. Taking up the idea, Moalem writes, "Every hairless mammal is aquatic or at least plays in the mud—think of hippos, elephants and the African warthog. But there aren't any hairless primates." (p. 198) Furthermore we have fat directly under our skin to help keep us warm just as aquatic mammals do. Also, Moalem notes, "the ability to survive on land and sea" gives us adaptive flexibility. If "chased by a leopard, the semiaquatic ape could dive into the water; chased by a crocodile, it could run into the forest." (p. 199)

These ideas are familiar but what I didn't know was that an aquatic past could have figured in our evolution toward bipedalism. "[S]tanding upright in water allowed...[aquatic apes] to venture into deeper water and still breathe, and the water helped to support their upper bodies, making it easier to support them on two feet." (p. 199)

This is an easy to read book, aimed at a general readership.

Nesse, Randolph M. and **George C. Williams** *Why We Get Sick: The New Science of Darwinian Medicine* (1994) ****
Readable introduction to the ideas of evolutionary medicine

This is a very readable book and an excellent introduction to a subject that has hitherto been sorely neglected. The main argument presented by Nesse and Williams is that disease must be understood from the perspective of evolutionary biology.

The authors begin by asking, "Why, in a body of such exquisite design, are there a thousand flaws and frailties that make us vulnerable to disease?" Through evidence and insights from evolutionary biology, the authors carefully give a detailed answer to this question, which might be summed up thus: The mechanism of evolution fits our bodies for reproduction, not for optimum health. Furthermore the mechanism is imperfect and subject to mutation. Additionally we are in competition with other organisms, e.g., viruses, bacteria, etc., that work toward their fitness, sometimes at our expensive (the parasite-host "arms race"). Noteworthy is the idea that natural selection cares little for the maintenance of the organism after the age of reproduction, and that sexual reproduction actually fosters mechanisms that increase the fitness of youth while neglecting the aged, leading to the phenomena of senescence and death.

Seeing disease from the viewpoint of evolution, the authors argue, helps us to understand disease and the mechanisms involved, which in turn can help us to fight disease. Allergy, for example, is a disease characterized by an over active immune system. Copious amounts of histamine are produced to fight off a few molecules of pollen. Why? The authors make the point that our immune systems operate on the principle that better an overreaction to something harmless than an under-reaction to a real threat. It's like jumping at the sight of a piece of rope lying on the ground. It's not a snake, but better this little harmless error than being too slow to get back from the real thing.

Some other interesting ideas: Fever has a purpose. It raises body temperature enough to interfere with the chemistry of some pathogens, thereby killing them. If we take medicines that reduce fever, are we prolonging our illness? In some cases, the authors answer, yes. If we take medicines that suppress coughs and sneezing can that also prolong our illness? Again the answer is in some cases, yes. The point is that in treating the symptoms of disease we need to make a distinction between which are genuine defensive mechanisms of our bodies and which are not. Some pathogens, for example, make us sneeze or cause diarrhea in order to better spread themselves to the next victim. The rabies virus makes a dog bite other animals in order to spread itself. But our bodies cause us to cough and sneeze primarily to expel pathogens.

The authors see some of our health problems as the result of genetic "quirks" or evolutionary hangovers. Dyslexia is an example. In the Environment of Evolutionary Adaptation back in the Stone Age, dyslexia was no problem because there were no books to read. Indeed, it might be that the dyslexic approach to some perception problems is better than the "normal" one, allowing a quicker and better understanding of the objects being viewed. Other genetic quirks include our predisposition to eat too much fat when available because in the EEA there was precious little fat to be had so it made sense to eat as much as possible when it was available. Something similar can be said of alcohol. Before agriculture, and especially before the process of distillation, a predisposition to alcoholism was no danger because there was very little alcohol to be had. These "quirks" are examples of disease caused by "novel environments," much of the modern world being a novel environment to our Stone Age bodies.

Nesse and Williams show that the modern environment, which requires a lot of close work from all of us, especially the reading of books, is the cause of the epidemic of myopia that modern humans experience. I would like to add that it is possible that myopia under some conditions could be adaptive. In the rainforest it would probably be better to see well close at hand than far away (the opposite of what would be valuable on the savannah). Also those people who concentrated on things small and up close might well identify and process food sources overlooked by others.

While this is an excellent book, gracefully written and full of valuable information and insight, it is now a little dated (copyright 1994), and some of the ideas need reworking in light of recent discoveries. For example, while the authors discuss the ill effects of too much fat and sugar in our diets, they say nothing about the carbohydrate intolerance that leads to obesity. This too can be seen as an evolutionary quirk since there were no cultivated fields of amber grain in the prehistory, and the grains that were available were small and required a lot of hand processing so that it was very difficult to overindulge. Consequently there was no need for natural selection to evolve a protection against eating too much. Also their discussion of heart disease and how it is the result of genetic factors and faulty diet fails to mention the idea that heart disease might be caused by a bacteria. (See for example, *Plague Time: How Stealth Infections Cause Cancers, Heart Disease, and other Deadly Ailments* (2000) by Paul W. Ewald.)

All things considered, though, this is a classic of evolutionary literature, nicely presented to a non-specialist, but educated public. Now if we can only get more doctors to read it!

Ridley, Matt *Genome: The Autobiography of a Species in 23 Chapters* (1999)
*****

A very effective format for Ridley

Matt Ridley proves here once again that he is a terrific writer. He has the facile style of a confident journalist and the wide knowledge of a budding scholar. He is learned without being stuffy. He proves too that he is a master of analogy and metaphor, understanding that we learn through comparison. I have the sense that he spent a fair amount of his free time looking for apt comparisons to illuminate the ideas of genetics for the general reader. Some examples: On page 276 where he describes the idea that there is a living thing with no DNA as "about as welcome in biology as Luther's principles in Rome." Or on page 241 talking about apoptosis, in which our cells are programmed to commit suicide: "the body is a totalitarian place." He even asserts on page 174 that we cannot hope to understand the process of embryotic development without "the handrail of analogy." My favorite is this from pages 247-248 where he is talking about gene therapy and an engineered retrovirus that doesn't work: "it lands at random...and often fails to get switched on; and the body's immune system, primed by the crack troops of infectious disease, does not miss a clumsy, home-made retrovirus."

Add a sharp wit and an infectious enthusiasm for understanding human behavior and one can see the reasons for his success as an interpreter of the biological sciences. In *Genome*, Ridley has found a structure and an approach that allows him to wax speculative and philosophical about matters of particular interest to him and to most people. The result is that the reader is treated to a lively mind at work trying to understand ourselves and this world we live in. He uses the 23 chapters, each emphasizing one aspect or our genetic makeup and each dedicated to one of our 23 pairs of chromosomes, to explore such matters as intelligence, instinct, the nature of disease, the effect of stress, the development of personality, memory, death and immortality, etc., and of course sex and—always an important question for Ridley—free will.

Some highlights:

The chapter on stress includes two startling assertions: One, that low status in the pecking order (instead of high cholesterol), lowers our resistance to microbes in our systems, and is the prime mover in making some of us more susceptible to heart attacks (p. 155); and two, that aggression is not caused by high testosterone levels but the other way around (p. 157). On page 171 he makes a similar assertion, namely that serotonin levels (as found in monkeys) are the result of dominate behavior, not the other way around, as has always been thought. These are exciting ideas since they suggest that we can improve our condition through our behavior (akin to "method acting," I suppose). Ridley's arguments strike me as convincing, but see for yourself.

In Chapter 21, he gives us a brief history of eugenics, noting, by the way, that during its heyday the name "Eugene" became popular in England. He spares eugenics practitioners and true believers not at all. He rips them up in true (and uncharacteristic) PC style, and then gets to his point. He likes eugenics but not the way it was practiced with the state coercing the individual. Instead Ridley would like (quoting James Watson on page 299) "to see genetic decisions put in the hands of users" instead of governments. He calls this "genetic screening" and cites the virtual elimination of cystic fibrosis from the Jewish population in the United States as a positive employment of screening from the private sector.

In Chapter 22 he tackles free will, beginning with a joke about there being a gene for free will. Clearly Ridley is in favor of free will, but reading between the lines one see that he knows he is on shaky scientific ground. He quotes the Oxford Dictionary of Philosophy on (David) "Hume's Fork: Either our actions are determined, in which case we are not responsible for them, or they are the result of random events, in which case we are not responsible for them." Ridley believes it is better to imagine that we are guided in our actions by our genes than by our conditioning. He sees nurture as being a more tyrannical dictator, if dictators we have, than our genes. This is not surprising since politically speaking Ridley hates the collective. He would love to have proof of the existence of free will since that is where his heart lies, but I hope that someday he will be comfortable with the understanding that whether we have free will or not (or whether "free will" is even a meaningful concept), one thing is clear: we have the *illusion* of free will, and that illusion is all compelling. Also, as Ridley notes, society must treat its members as having the ability to make free choices or the whole system of law collapses.

Perhaps the most amazing feat of our genome is the one Ridley writes about in Chapter 12, that of "Self-Assembly." To me that is the really stupefying trick of our genes, to assemble themselves from the code. The twists and turns of such an enormously complex undertaking is, to me, as remote from our understanding and experience as the eleven dimensions of super string theory.

Other popular writers on science looking for the secret of Matt Ridley's success should note that he gives the reader value both in terms of knowledge and entertainment. He works hard at helping the reader to understand what he is saying. Even though I sometimes disagree with him, I always learn something new and interesting from reading his books.

Ryan, Frank *Virolution* (2009) *****
The starling (and scary) role played by viruses in biological evolution

A major thesis of this amazing book is that plants and animals including most significantly humans co-evolve with viruses. The term "virolution," presumably coined by Dr. Ryan, who is both a physician and an evolutionary biologist, comes from the words "virus" and "evolution" but also suggesting the word "revolution." The idea is that instead of being merely agents of pathology, viruses can also work together with their host to help it survive. Ryan gives the example of grey squirrels imported from America invading the territory of red squirrels in Britain. He writes:

"At first naturalists assumed that the grey squirrel was winning the survival battle because it was larger and more aggressive than the native counterpart, but now we know that the grey squirrel is carrying a squirrel pox virus that causes no disease symptoms in its symbiotic partner but appears to be lethal to the red squirrel." (p. 96)

In other words what we have here is war by an organism's own viral pathogens! Survival of the fittest may include carrying around lethal viruses that can wipe out your ecological competition. Ryan notes "We believe that HIV-1, the main virus of AIDS, was transferred to people from a specific group of chimpanzees. We also know that, in chimpanzees, HIV-1 grows freely and reproduces in their internal organs and tissues, but it causes no evidence of disease." (p. 86)

So what apparently happened is some bush meat eaters shot some chimps, ate and/or sold the meat and humans got the virus. Revenge of the dead chimp! Well, perhaps. But look at it this way. Imagine humans in prehistory or even humans a few centuries ago in the Congo jungle looking to take over some chimp territory. After some close contact, the virus jumps from the chimps to humans and the humans die. Survival of the fittest!

Ryan refers to this as an example of "aggressive symbiosis," and this is how it works in general: two similar species occupying similar ecological niches come into contact. Which is to prevail? One carries a virus like a loaded gun in its

78

tissues. The virus jumps to the other species and typically is extraordinarily virulent and kills them. Or perhaps there is a dueling of viruses, one from each species. At some point the only survivors are those with immunity to the viruses.

Ryan makes a further point with this example (quoting Max Essex on the deliberate use of a myxomatosis virus to kill rabbits in Australia): "The...virus killed...some 99.8% of the rabbits. But then two things happened. Number one - within four years, the resistant minority grew so you had a different population of disease-resistant rabbits... And number two - the myxomatosis virus that remained [as a persistent infection in the rabbits] was less virulent, so I think there is crystal-clear evidence that both the host and the virus attenuated themselves for optimal survival in that situation." Furthermore (and this brings us back to the previous point), any new rabbits brought in would be at a disadvantage because they would have no immunity to the virus and the surviving rabbits would. (pp. 87-88)

In other words looked at from an evolutionary perspective, host and virus worked together in a mutualistic symbiosis. In my mind this raises the question, what really did happen to the Neanderthal? We do know what happened to the natives of the Americas when they came into contact with the smallpox virus carried by the Europeans. Could a virus from *homo sapiens* have wiped out the Neanderthal, or at least helped humans become the sole hominid survivors?

In the largest sense, this idea of host and virus working together would seem to be more powerful than any kind of sharp tooth and massive claw in the struggle for survival. The old idea of survival of the fittest must now be seen in a different light. I have said for many years that "everything works toward an ecology" and "everything works toward a symbiosis," meaning that in a typical environment, if one species is able to work together with another, they may enjoy an advantage over rivals. Consequently, those species that are able to form symbiotic relationships are the ones more likely to survive. What this means for evolutionary theory, as Dr. Ryan has pointed out, is that symbiosis is a much more important part of evolutionary biology than has previously been thought. My guess is that the revolution begun by Lynn Margulis, who first saw the eukaryotic cell as a mutualistic development from parasitic relationships, will be accelerated by the work of Ryan and others to the point where the prevailing view from evolutionists will be that it is cooperation rather than competition that most characterizes fitness.

And that is what makes this book so important. It signals a great shift in our understanding of how evolution works.

But that is not all. Ryan shows that the so-called "junk DNA" in genomes is anything but. Much of it is viral ("endogenous retroviruses") and it is there as evidence that humans and pre-humans went through many periods of aggressive symbiosis including the horrid plague stage. We now see that plagues, from an evolutionary perspective, are common and part of how the evolutionary process formed us. Furthermore Ryan writes about how viral genes can help with the development of the embryo in the womb. In other words, viral DNA in part directs the protein building that makes for human beings, and indeed for many forms of life.

In the latter parts of the book Ryan explores the role of viruses in autoimmune diseases and cancer. He also considers the role of hybridization in evolutionary change and that of epigenetics. Particularly interesting is the work of Eva Jablonka and Marion J. Lamb that suggests that "new species might arise through the inheritance of *acquired epigenetic changes*," causing Ryan to remark, "they were resurrecting the long-discredited spectre of Lamarckian evolution." (p. 312)

The book is dense, difficult and perhaps revolutionary in scope.

Shnayerson, Michael and **Mark J. Plotkin** *The Killers Within: The Deadly Rise of Drug-Resistant Bacteria* (2002) *****
The frightening return of infectious disease

This book is scary. According to ethnobotanist Mark J. Plotkin and longtime *Vanity Fair* contributing editor Michael Shnayerson, the golden age of antibiotics that began with penicillin, a time when it was generally thought that infectious diseases were under control and largely a menace of the past, is over. Our naiveté and our arrogance in imagining that we had just about defeated the bugs and could move on to other more pressing public health concerns came to an end in the nineties as one after another of the major human borne bacteria became resistant to our drugs. Through the exchange of DNA, that immunity has been transferred to other bacteria so that, as this book went to press just a few months ago, infectious diseases caused by bacteria are once again a major threat to humans everywhere in the world.

What happened? As the authors explain there are three main problems, (1) the overuse of antibiotics by the medical profession, (2) the misuse of antibiotics as growth enhancers in the meat and poultry industry, and (3) the failure of hospital personnel to follow CDC guidelines on hygiene, especially simply washing their hands.

(1) Too many doctors, either through ignorance or a desire to please their demanding patients, have over-prescribed antibiotics for routine infections, and in some cases actually prescribed antibiotics for viral infections (for which they are useless) "just in case" the patient also gets a bacterial infection. The result of this massive overuse of antibiotics is to give the bugs countless trillions of generational opportunities to evolve defenses against the antibiotic, leading to the antibiotic becoming useless.

(2) Tons of antibiotics—"24.6 million pounds a year," see p. 123—are routinely added to animal feed by the meat and poultry industry to promote growth so that their products will get fatter faster. What has happened is that the bugs have grown resistant to the antibiotics while transferring that immunity to bacteria living in, on and around humans. Even the use of an "analogue" antibiotic such as growth promoter virginiamycin can promote changes in bacteria that make them resistant to the antibiotic Synercid (e.g., see pages 115, 119 and 285). As the authors chronicle, this is a serious problem fraught with angry political battles as the meat and poultry people fight to maintain their profit margins while the disease control people fight to restrict the use of growth promoters.

(3) Surprisingly enough the authors report (see page 282 and elsewhere) that there is cynicism among some hospital personnel about the effectiveness of washing their hands and a belief that hygiene won't stop the proliferation of the bugs. The result is that hospitals have become very dangerous places. Most of the drug-resistant bacteria developed their resistance in hospitals. Most (or all) of them are endemic to the hospital environment. If you have to go to a hospital for any reason you are taking a chance of contacting a drug-

resistant bug. Heaven help you if you have a compromised immune system, or if you are an infant or an elderly person.

How bad is the situation? According to the authors on pages 278-279 the high cost of developing new drugs (average "$802 million") and the fact that "return on investment from producing an antibiotic that might be used by a patient for less than a week versus return from a drug for a chronic condition that a patient might take daily for fifty years" is persuading big pharma to downsize the antibiotic end of the business. (See also page 94.) The authors ask the question, where are new drugs coming from? and answer that the "great glittering prospect was genomics." But "reality" has "sunk in." (p. 280) Drugs to fight bacteria developed from DNA manipulation "might take even longer to reach the market" than those previously developed. (p. 281)
The authors also touch on the possible use of drug-resistant bacteria as a bioterrorist weapon.

What does all this mean for you and me? It means that should we or our loved ones get a life-threatening bacterial infection, it's possible there won't be an antibiotic around that works. In effect, we might find ourselves back in the days before penicillin (the first really effective antibiotic, and one of the greatest of all medical miracles) when millions of people routinely died from staph, strep, TB and other bacterial infections. As Shnayerson and Plotkin report, right now there are strains of bacteria, including *Staphylococcus aureus* (the golden-globed bacteria pictured on the cover), *Streptococcus pneumoniae*, and the *Mycobacterium* that causes tuberculosis, that are immune to almost every antibiotic in use. There is even a strain of *Enterococcus* that is resistant to *every* antibiotic in use.

The authors do offer some hope. They report on the promising use of bacteriophages (viruses that invade and destroy bacteria)—see the very interesting Chapter 14, "Bacteria Busters." They present the idea of a more vigorously controlled use of antibiotics. If we prohibit their use as growth promoters and use them sparingly in an ordered sequence, perhaps bacteria would not have time to gain immunity and/or would lose it after the antibiotic is no longer in use. As pointed out on e.g., page 183, resistant bacteria are "encumbered" by an "extra chunk of DNA" that gives their non-resistant brethren an "ecological advantage" in an environment that doesn't contain the antibiotic. Additionally, the authors report the theory of population biologist Richard Levins who believes that if antibiotics are "saved for the most severe cases...then natural selection would favor the pathogens that produced the milder symptoms." (Explained on page 287).

Bottom line: this is a fascinating, scary and state of the art report on the pathogen wars written in a readable manner sure to interest not only the general public at which it is aimed but professionals as well.

Sykes, Bryan *The Seven Daughters of Eve: The Science that Reveals Our Genetic Ancestry* (2001) *****
How genetic knowledge is rewriting the prehistory

This is a popular book of scientific discovery written in an affecting and engaging style by a geneticist who has the all too rare gift of writing extremely readable prose.

Professor Bryan Sykes draws the reader into his story as easily as a best-selling novelist. And this is just the "science" part of the book which lasts for fourteen chapters. Then come the fictional chapters about the seven daughters and their imagined stories, so touching and so full of the very human struggle to survive in the prehistory that I could not read them without misting up. (But then I tend to the sentimental.)

Sykes begins with the story of how he was able to identify a living descendant of the five-thousand year old "ice man" found in northern Italy in 1994 by comparing mitochondrial DNA sequences. Mitochondrial DNA is contained only in egg cells (thus, "Eve" and her daughters), not in sperm cells, and transmitted without recombination so that the changes are all the result of mutations that occur at a predictable rate over time. Then he tells the story of how the bodies of the murdered Romanovs, the last of the Russian Tsarist families, were identified through DNA fingerprinting. Both of these stories are more about media events and ventures in forensics than original scientific work. But then comes the story of where the Pacific Islanders originated.

When I was young I read the engaging story of Thor Heyerdahl in his book *Kon-Tiki* in which he attempted to prove that the Polynesians originated in the Americas by sailing west into the Pacific. This beguiling theory is demolished once and for all by the DNA evidence that Sykes presents. He shows that the Polynesians were originally from Southeast Asia and made all their great ocean discoveries by sailing against the prevailing winds, going east toward the Americas.

Sykes notes that because this was the prevailing scientific opinion his work met with mostly agreement. However when he and other geneticists were able to show that the current population of Europe is mainly descended from the original hunters and gathers that lived there prior to the arrival of the farmers who brought agriculture from the Middle East roughly ten thousand years ago, they ran into resistance. The prevailing scientific opinion was that the farmers overwhelmed the hunters and that most of today's Europeans are descended from those farmers. Sykes relates the story of the scientific controversy and how the genetic proof finally prevailed against entrenched opinion. Incidentally, to me the intriguing thing about this discovery is the question, not addressed in the book: What, if any, conclusions can we draw from the fact that 80% of our European genes came from hunters and gathers and only 20% from Middle Eastern farmers?

There is also the story of the "Cheddar Man" and how Sykes learned to extract DNA from the bones of people dead tens of thousands of years. Finally there is his argument for all people of European descent coming from just seven women who lived ten thousand to forty thousand years ago, the so-called, "Seven Daughters of Eve." (World-wide Sykes identifies 33 "daughters of Eve.")

To round out the book, Sykes writes an imaginative chapter about each one of the seven daughters. Here is where some readers are displeased, claiming that Sykes's imaginings are unscientific and even slanted. One reader complained about the men out hunting and the women remaining behind in caves as a kind of stereotype that has been overcome. But remember Sykes is writing in six cases out of seven about European peoples who made their living primarily from hunting during the ice ages, not from gathering. Think about how much "gathering" the Inuit do and you can see why he emphasized hunting. In the seventh case, about Jasmine, whom he sees as being from the birthplace of agriculture in modern Syria, his story is different. Indeed he has Jasmine and her non-hunting mate inventing agriculture! I might also point out for those who skimmed the "daughters of Eve" chapters, that he also has a woman playing a major part in the invention of water-going craft.

If I were to criticize this book I would say he was too generous in his depiction of human beings in the prehistory. He describes their lives as hunters and gathers, their hardships and their short and difficult lives with an emphasis on their humanity and how that helped them to survive. He downplays any part humans may have had in the extinction of the Neanderthal. He relates no rapes or murders or tribal wars, and de-emphasizes tribal sexism. He shows the beginnings of trade and cooperation. The result is so warm and touching I'm surprised that Stephen Spielberg hasn't taken out an option on the book. (Maybe he has!)

Finally, this is not an academic tome. It is a popular science book meant for educated lay persons. There are no learned academics writing glowing blurbs on the cover. Most academics would be afraid to write a book like this because of the imaginative chapters which are quasi-scientific and can be so easily criticized.

In short Professor Sykes is a tremendously engaging writer (with guts) who happens to be a world-class scientist. His goal was to communicate something about what he has learned to a wide readership, and I think he did a good job. If you can read this book without feeling better about humanity, maybe you should read it again.

Wakeford, Tom *Liaisons of Life: From Hornworts to Hippos, How the Unassuming Microbe Has Driven Evolution* (2001) *****
We and the microbes are one

This book is about symbiosis and how prevalent it is. It is also about how politicized the concept has been historically. From the experience of nineteenth-century biologist and illustrator Beatrix Potter whose identification of lichen as symbionts went against the established dogma as filtered through the ideas of Pasteur, to "anti-communist" biology as practiced by some Western scientists who saw symbiosis as supporting the collective, it is amazing how purely political ideas successfully censored the scientific. Symbiosis has even been thought of as "feminine" and contrary to the noble interpretation of Darwinism as the survival of the fittest.

But Wakeford is able (after a fashion) to go beyond the politics and demonstrate in a most convincing manner that the symbiotic way of life is vastly more important and enormously more widespread than is usually imagined. Most of us know that legumes work symbiotically with rhizobia bacteria to fix nitrogen in the soil so that it is available to the plant, but what surprised me is to learn that 90 percent of plants host mycorrhizal fungi (p. 167) and are therefore symbionts. As Wakeford asks on the same page, "Can we continue to simply call them plants without acknowledging their fungal dimension? Is a cow an animal or a microbial fermentation vessel, when without the microbes, the cow would not exist?"

Good questions, and indeed, what about humans who have microbes in our guts that help us to digest our food? Are we in symbiosis with those microbes? Without the beneficial bacteria in our guts, the harmful bacteria would run rampant and we would be led to disease. Ants are not merely ants, they are farmers who harvest fungi gardens. They and the fungi are in symbiosis, living together, dependent upon one another for their survival. And what about termites, creatures who harbor microbes to digest the wood they eat? The broad, general message of this book is that cooperation between species is at least as important in evolution as is competition.

Reading this made me think that perhaps the idea of competition in evolution is merely an anthropomorphic delusion. Certainly Wakeford shows that our notions about parasites and who is feeding on whom, may be in error. He writes, "Rather than discrete categories, the terms *mutualist, parasite*, and *pathogen* are better seen as fuzzy points on a continuum, along the length of which an association between two organisms may fluctuate. For many associations, the point they occupy on this continuum is as difficult to assess as it is to say who gains more...in a marriage between two human partners." (p. 184)

There is an old saying, that I got from somewhere years ago. It is, "Everything works toward a symbiosis." This book not only supports that idea, it even, taken further, supports the idea of Gaia, namely that all the living creatures on this planet form a single organism. I don't necessary believe this, the "strong" Gaia hypothesis, but I think the distinction between a planet that harbors organisms and a planet that is itself part organism, may be more a semantic distinction than anything else.

Because of all we have learned about microbial life in recent decades, it is becoming clearer and clearer that no organism is an island, and indeed, all of life is in symbiosis with the microorganisms that constitute the largest, most viable life form on this planet. Realizing this while reading Wakeford's fascinating arguments, I had a thought: the little green men from outer space are probably symbionts themselves, but more fully realized ones, like lichen, part "animal" and part "plant," deriving their energy directly through photosynthesis. And suddenly I had a vision of beings all seated as in meditation, taking a break to open the top of their "heads," filled not with brains, but with cells capable of turning light into nourishment. How primitive and clumsy we might appear by comparison!

Chapter Four

Evolutionary Psychology

Evolutionary psychology has run into some of the same problems with the general public that plagued Darwin in his day. Even some biologists on the left have found its discoveries hard to swallow. The main problem is that evolutionary psychology posits that our behaviors are often genetic adaptations honed in the so-called Environment of Evolutionary Adaptation of long ago and are not culturally derived and indeed resist modification by education. People on the left like to think that nurture in general, education and the force of society are more important forces in shaping our behavior. People on the right like to think that for the most part we are born that way. Consequently evolutionary psychology becomes a political football kicked and tossed around by those whose biology is influenced by their political persuasions. Academic evolutionary biologists and other professionals are not immune.

On the left we find for example, Richard Lewontin and Steven Rose, and to a lesser extent, Stephen Jay Gould; and on the right we have Edward O. Wilson, Richard Dawkins and the journalist Matt Ridley, among others. My position, even though I tend to the left politically, is to side with the nature side of this pseudo nature vs. nurture controversy. I think our behaviors are hardwired in to an extent that would surprise the average person.

At any rate here are reviews of 29 books on evo psych, most of which put forward the proposition that we behave the way we do because such behaviors proved valuable and adaptive for our ancestors in the Pleistocene.

The key to understanding human behavior in my opinion requires knowledge not only of evolutionary psychology but of cognitive psychology and neuroscience, and of course a lot of experience.

Allman, William F. *The Stone Age Present: How Evolution Has Shaped Modern Life—From Sex, Violence, and Language to Emotions, Morals, and Communities* (1995) ****
Good, but keeps the rose-colored glasses on

The point implied in the title is a good one: we are stone age animals living in an electronic jungle. The Environment of Evolutionary Adaptation, which was on the savannas of Africa, disappeared for most of us long ago; but genetically and phenotypically speaking we have changed very little. Thus the first four words of the title are beguiling; the rest after the colon, I suspect, was something formulated by a committee of book biz editors trying to spice up the presentation.

This is evolutionary psychology written by a journalist, readable with some worthwhile insights. It should be compared to Richard Wright's *The Moral Animal* (1994) and Matt Ridley's *The Red Queen* (1993) from the same time period. This is a comparison that could be extended to other books on evolutionary psychology, including anthropologist Marvin Harris's *Our Kind* (1989): sociobiologist Edward O. Wilson's earlier, *On Human Nature* (1978); Robert Jay Russell's *The Lemur's Legacy* (1993); Richard Wrangham's *Demonic Males* (1996), etc. Incidentally Amazon has all these books and others, so you might want to do a little comparison shopping. What one notices is that Allman's book is, relatively speaking, a feel-good, sanitized narrative. Our stone age ancestors did not kill a cow and cut up its carcass and distribute it to others in order to enhance their power and prestige and to gain reproductive favors, as most "observers" would have it; but, according to Allman, to share "with friends and neighbors" and "courting lovers." It is amazing what a difference terminology can make. Allman almost allows us to embrace evolutionary psychology and its rather unflattering insights and keep the rose-colored glasses on. The tone is positive and reasoned. The book is also as politically correct, although not as annoyingly PC as Wrangham's *Demonic Males*.

I should mention that one of the major themes in this book and in recent evolutionary psychology is that our brains grew big and smart to deal with our complex social lives. This is the current wisdom. Well, as Satchel Paige said, "The social ramble ain't restful," and as I've always said, socializing is a lot of work. Yes, I think this really does explain how our brains got to be so big. We needed to be really smart to outsmart the other guy. We needed to be smart to juggle all those intrigues, social, political and sexual. I like the way this insight fits with the female's abhorrence of nerds: the fact of the matter is, not being social is also not being smart! So there, nerds!

Like Harris, Allman does not see civilization or the rise of agriculture as necessarily a good thing for the average Joe. And he is firm in discounting the idea that human beings represent "progress" on the evolutionary scale. Interestingly, Allman reports extensively from Robert Axelrod's work on cooperation in an attempt to make us look like good guys. Axelrod is the guy who devised the computer models testing the prisoner's dilemma and held the competition that revealed the now well-known and celebrated "Tit for Tat" strategy that won it (initially cooperate and then act toward the other as that other has acted toward you: tit for tat). Tit for tat also appeared in Wright's *The Moral Animal* and in Ridley's *The Origins of Virtue* and elsewhere. I think Axelrod might have had a press agent. At any rate, tit for tat is now seen as needing a random and forgiving variation in order to defeat various other strategies, including ruthlessly non-cooperative ones.

This is a pretty book, originally from Simon & Schuster, very well edited and proofed (thank you!).

Boyer, Pascal *Religion Explained: The Evolutionary Origins of Religious Thought* (2001) ***
Misses the point

Boyer sees religion as a by-product of the way our minds have evolved. His "explanation"—laboriously presented in a most excruciatingly detailed manner—left this reader exhausted and a little annoyed. Much would have been gained had the text been reduced by perhaps two thirds. Although Boyer writes in a clear manner, the tedious qualifications and the needless repetitions make the book exasperating to read.

But that's just the minor problem. The major problem is that after all these words, Boyer does not really explain much at all. Clearly religion of one form or another is found in virtually all human societies. Consequently it doesn't take a very sophisticated deduction to conclude that we believe the things we believe because our minds work that way. Religion is part of human nature, hard-wired to some very real extent in our brains similar to the way grammar is. What needs explaining is how religion is adaptive. If it didn't somehow increase our ability to survive and reproduce—that is, make us more fit—it would not be universal.

This is the key that Boyer marches around, hovers over, and, alas, misses. As famed biologist Edward O. Wilson wrote, "When the gods are served, the Darwinian fitness of the members of the tribe is the ultimate if unrecognized beneficiary." (*On Human Nature* (1978) p. 184)

But just how does religion increase the fitness of the members of the tribe? By making them accept their lot on earth because they will get their reward in the hereafter? As Boyer points out, this can't be the answer—at least not the entire answer—since some religions don't have a hereafter. How about increasing the coalition among members of the tribe thereby increasing cooperation and mutual trust? I think this is on the right track. If the tribe works together toward a common goal, the tribe will be more effective in dealing with the environment, and the tribe will increase. But what is the most important and demanding aspect of the human tribal environment? Other tribes!

This brings us to what I think Boyer missed entirely: the war system. Religion, because it persuades people to believe in things greater than themselves, facilitates the kind of fearlessness that is most effective in killing members of the other tribe. A tribe that has a ferocious leader who is followed as one might follow a god, or one of god's representatives, is a more effective fighting unit than a tribe that doesn't have that kind of cohesiveness. If a belief system can get the young males of the tribe to lay down their lives for the good of the tribe, that tribe will take over the rich valley, the land of milk and honey, and the less fit tribe will go into the mountains or perish.

If the leader of the tribe can get the tribe to see that a victory over the enemy is God's will or that God or the spirits or the angels are on the side of the chosen tribe, so much the better. To make this work people have to be able to believe in things not seen or understood, things that go bump in the night, things mysterious, frightening, things brought forth by the shaman amid smoke and ritual.

But as Boyer points out, no single explanation for religion is adequate. In religion we also find the beginnings—paradoxically—of science. When the rains didn't come and the grain grasses didn't grow and the animals became few, the people asked why and wanted to know what they could do. Religion supplied the answer. Throw the sheep bones and know which way to go. By happenstance the tribe wandered in the right direction and this was remembered. Sacrifice an animal to the gods and the gods will cause the rains to return. (And if the rains don't return, you did it wrong.)

This is sympathetic magic, surprisingly not mentioned specifically by Boyer, although he treats superstition at some great length. Sympathetic magic is part of almost all religions in the form of ritual and prayer. It was but a step or two (giant steps of course) from throwing salt over one's left shoulder to broadcasting plant seeds over the ground. Sympathetic magic which is at the heart of religion became, after many a moon, science.

Although Professor Boyer admirably attempts to account for religion from an evolutionary point of view using an anthropologist's eye, I am afraid that he got lost in the thickets and missed the pure essence of his subject matter. I suggest he read some Edward O. Wilson and Marvin Harris (both absent from his bibliography). Harris shows how religious beliefs work to support adaptive behaviors (e.g., not eating cows in India) while Wilson will give the reader a good understanding of human psychology from an evolutionary point of view.

Budiansky, Stephen *If a Lion Could Talk: Animal Intelligence and the Evolution of Consciousness* (1998) ****
Very much worthwhile, but contentious

This is a very slippery book on a very slippery subject. What Stephen Budiansky is trying to do is demonstrate from his reading of the literature, including experiments published in peer-reviewed journals, that there is a distinction to be made between the minds of humans and all other animals. Budiansky seems not to believe that intelligence and consciousness are matters of degree, but matters of threshold. Following philosopher Daniel Dennett he attributes this nearly absolute difference between us and them to our ability to use symbolic language.

The reason the subject is so slippery is that an adequate definition of both intelligence and consciousness is lacking. The reason the book is contentious naturally follows from this, but additionally Budiansky seems to have an agenda or, call it a thesis. He writes: "Consciousness is a wonderful gift and a wonderful curse that, all the evidence suggests, is not in the realm of the sentient experiences of other creatures." (p. 194)

How true or not his statement may be really depends on the definition of consciousness. Unfortunately Budiansky does not give one, and so all his conclusions about the differences in consciousness between humans and other creatures are murky at best. The closest he comes to a definition is on page 193 where he asserts that "...language is so intimately tied to consciousness that the two seem inseparable." Using this "definition" it is only a matter of demonstrating that animals do not have language in order to demonstrate that they don't have consciousness.

However even in this I don't think Budiansky is successful. Much of the book is given over to showing how so many experiments using chimps and monkeys, pigeons and dogs, etc., that seem to demonstrate that language use by animals is just signaling. This position is well known. The argument is that humans are the only animals with grammatical, syntactical and symbolic ability built into their brains. Other animals cannot construct sentences because they have no syntax. They have no "theory of mind" because they cannot think symbolically.

But this is not proven, as Budiansky acknowledges. What is obvious is that whatever language ability other animals have is rudimentary compared to that of humans. And almost everyone would agree that the consciousness demonstrated by animals varies considerably. By the way, here's a quick definition of consciousness: awareness including self-awareness, identify, and experience or sensation: the feelings we get when we experience the world, like the taste of ice cream. A lot of confusion results because when people talk about consciousness, one person may have in mind "awareness," while another may be talking about "self-awareness" only, or about "self-identity," while another may be talking about what it feels like to be in love. Awareness includes past, present and future events, and places here and elsewhere. We are very good at all of this, whereas other creatures are apparently not so good at anything other than the here and now. Because of our extended awareness, people like Budiansky are persuaded that we are on a consciousness level above other animals that should be recognized as different in kind.

Notice, by the way, that the idea that consciousness depends on language is by this definition obviously false. Sentient beings can be aware of many things without using language. Also there are different kinds of languages. Budiansky is talking about the kind of language that linguists study, the kind of language that Norm Chomsky analyzed to come up with his discovery that syntax is innate. But mathematics is a language, and when mathematicians are thinking about equations, they are conscious to the same extent that I am when I am thinking about how to put an idea into a sentence. Ditto for chess players and musicians. The languages that humans use are of one kind. We do not yet understand the language the whales and dolphins speak.

What I don't like about Budiansky's insistence on a difference in kind is that when you stop to think about it, such a difference would be surprising since all life forms on this planet evolved from a single ancient ancestor—unless of course you believe in a divine and separate creation.

Some other points at issue:

Budiansky wants to debunk the idea that animals are "worthy of special consideration" because their "behavior resembles" that of humans (see, e.g., p. xiii). I agree. We should appreciate other living things for what they are and not for how much they resemble us.

Consider the example of a chimpanzee holding out her hand to another in an appeasement gesture only to attack the other when he got near. Budiansky writes that a "theory of mind" interpretation would be that the tricky female knew the male would be misled in approaching and took advantage. But the "behaviorist spoilsport" interpretation is that the female had done this in the past and it worked and so did it again without recourse to reading the other's mind. (p. 182) This example illustrates just how difficult it is to say what is going on in another's mind. Personally I think the notion of a "theory of mind" should stay in the philosophy department except as demonstrated empirically.

One of the things that Budiansky makes clear is why some animals cry out when a predator appears. (See Chapter 6, "Speak!") Such calls seem altruistic to the point of being impossible from an evolutionary perspective; however

Budiansky argues that such cries actually help the crier because their pitch either fools the attacking hawk so that it looks in the wrong direction, or the calls bring out other victims who go running about, thereby confusing the attacker or giving the attacker targets other than the crier.

Another nice thing that Budiansky does is show in sharp detail that the language accomplishments of chimpanzees and gorillas in some famous studies reveal not so much a human-like ability, but demonstrate the great gulf that exists between our use of language and theirs, which is not the kind of truth some people want to read.

Buss, David M. *The Dangerous Passion: Why Jealousy Is as Necessary as Love and Sex* (2000) ****
"Necessary" from the POV of the genes...

Jealousy exists, like love and sex, to help propagate an individual's genes. It is a mechanism of the species to help insure for males paternity, and for females that their offspring receive the benefit of male protection, support and guidance. Jealousy is not "necessary" (as the subtitle disinformationally suggests) in the same sense that sex per se is necessary; nor is it an emotion, like love, that we might want to retain, had we our druthers. Jealousy is the emotional downside of the sexual/reproductive strategies employed by humans. It is "necessary" in the same sense (although not to the same degree) that pain is necessary. Furthermore, in the environment we now find ourselves, as opposed to the prehistoric savannahs in which the mechanism of jealousy proved adaptive, it is unnecessary, and something we might want to understand and come to grips with in an attempt to lessen its hold on us.

But what this book is really about is infidelity, how and why it occurs, and what can be done to forestall it. In this context, jealousy (not *envy* which is directed at somebody who has something we want) is seen as an adaptive mechanism to protect the individual against a straying partner, either through heightened awareness or through inducing threats of reprisal, or through actual punishment of the infidel. Buss, a psychologist and author of the college text, *Evolutionary Psychology: The New Science of the Mind*, uses case histories from our culture and others and the results of personality inventories laced with humor to illustrate how the experience of jealousy leads to "mate guarding" and "mate retention tactics" that help the individual secure his or her position in the "mating market." As such jealousy is seen as a "signal" to both one's self (awakening one to the imminent danger of infidelity) and to one's partner (as a warning that one is on to the other's tricks). Consequently, Buss defines jealousy (p. 196) as "an adaptive signal of an impending threat to a primary love relationship." Included in this view is the understanding that infidelity, painful as it is, is a normal human behavior practiced by "as many as half of all married individuals."

The style here is easy and accessible to a wide range of readers. The material is light-hearted (inasmuch as such a serious subject can be) but without any pasting-over of the dangers of jealousy. Underpinning the exposition is a thorough knowledge of human sexuality as derived from biology and evolutionary psychology. Buss not only knows what he is talking about, but imparts the information in a manner that, chapter by chapter, leads the reader to a deep and satisfying understanding of infidelity and the mechanism of jealousy.

Along the way we learn some unsettling facts. For example, marital happiness has no effect on the instance of male infidelity. "In fact, 56 percent of the men who were having affairs judged their marriage to be very happy." (p. 146) Or that women pursue a sexual strategy including a "desire to stray" that "exists today solely because that's what benefitted ancestral women" (p. 159). We also learn which type of personality is likely to stray (pp. 148-151) and that the more attractive partners ("those...higher in mate value") are more likely to cheat (p. 143). Also interesting is the semi-obvious observation that women can attract a higher-ranked male on a one-night stand than as a husband (and so might), and that men will stoop to lower-ranked females for pure sex than those they choose for wives.

Buss devotes the last two chapters to coping mechanisms. He concludes with the fine observation that "knowledge...of our dangerous passions...will, in some small measure, give us the emotional wisdom to deal with them." This observation is what evolutionary psychology is all about, and why it is the emergent psychology of the twenty-first century.

Best joke (p. 185): At a therapist's gathering with a straying husband, his wife and the other woman, the wife informs the affairee that she is still sleeping with her husband, and that he has lied to both of them. "The affairee felt betrayed and stalked out, saying...that all men betray their wives, but only a real asshole would betray his girlfriend." Buss adds, "Therapy was unsuccessful in this case."

Dabbs, James McBride with **Mary Godwin Dabbs**. *Heroes, Rogues, and Lovers: Testosterone and Behavior* (2000) ****
Readable report on the latest research

Males commit violent acts at a rate much greater than women. The vast majority of people in prison are males. One of the reasons is they have more testosterone pumping through their veins than women. Testosterone makes people take chances. It makes them more interested in sex and more aggressive. It makes them into "heroes, rogues and lovers," to quote the title of this interesting book. Testosterone tends to affect low socioeconomic status males more than high status males, and the effects of testosterone can be mitigated by learning. Women also produce testosterone, but at lower levels than men; however, what they do produce affects them more. Women are attracted to high testosterone males, but do not necessarily marry them. Women select males and thereby create the males that exist. We inherit our testosterone levels, and testosterone comes before rambunctiousness, not the other way around. (This last from pages 87-88.)

These are some of the facts gleaned from the research of Professor Dabbs, who is the head of the Social/Cognitive Psychology Program at Georgia State University. This book is a report on that research presented with examples, allusions and references to literature and the popular culture, leading to an easy read. Dabbs, along with his collaborator, his wife, Mary, "a former publicist with several feminist organizations," allows us to see the world through testosterone-shaded glasses, but without prejudice. Their report is balanced and fair. They give us the downside of testosterone and the upside, as implied in their title. The fact that theirs is the first popular full-length book (that I know of) devoted exclusively to the phenomenon of testosterone is the result of fairly recent technology that allows the measurement of testosterone levels from saliva samples. Previously, blood had to be used. Since most people are more willing to spit than to allow blood to be taken from their bodies, this technique opened up new possibilities for research, and Dabbs, who apparently has a fair amount of testosterone still pumping through his veins, got there first.

There are charts and graphs showing testosterone levels by occupation. Construction workers, actors, football players, con men (!), blue collar workers, etc., predictably are high in testosterone while clerical workers and clerics, counselors and farmers, etc., are low. Lawyers tend to be high, with trial lawyers and especially flamboyant defense lawyers the highest, with research lawyers the lowest. Relatively high testosterone levels correlate with masculine traits such as muscle strength, spatial ability, narrow-focused thinking, combativeness, while lower levels correlate with feminine traits such as sociability, more generalized thinking, verbal ability, cooperation, etc. Men tend to leap to action, while women tend to think about it first. Higher testosterone does not correlate with high economic status since our society rewards thoughtfulness, patience, and cooperation as well as hard work and being assertive. High testosterone males die younger but have more sex. This too is a predicable finding since it is a type of evolutionary strategy. Testosterone, in fact, might be seen as the chemical form of aggressiveness. Aggressiveness is getting there first with the most. It's a kind of strategy that often works. But there are problems as well as rewards in aggressiveness. First, it's costly; you use more energy. Second, you're not as sure in your actions so you make more mistakes, which is dangerous Third, you incite aggressiveness on the part of others, and that too can be dangerous. Fourth, sometimes getting there first may lead to no advantage. Finally, you can be only so aggressive. Aggressiveness leads to an "arms war." If aggressiveness is rewarded—and it is in a passive world—then everybody tends to become more aggressive until nobody has an advantage; in fact the passive now have the advantage because they live longer, etc., leading to the selection of more passive creatures, creating an environment effectively exploited by the more aggressive, leading to...the arms race cycle.

Some interesting quotes from the text:

"Wife abuse" tends to increase "in the Washington, D.C., area after the Redskins win their football games." (p. 92)

"To some men, a good relationship allows them to strut while their wives admire them." (p. 111)

"...[W]omen know in their secret hearts that men who won't kill for them are useless." (p. 61. Dabbs is paraphrasing from Cormac McCarthy's novel, *The Crossing*.)

"...[C]avewomen had to have resources and protection for their young, and so in courtship and mating, they favored dominant and powerful suitors...<Cavewoman> values persist today...Money is associated with power...women want men with <good financial prospects>." (p. 113)

"Senator William Proxmire once denounced two of my colleagues for looking at love scientifically, saying that love was a mystery, not a science, and he wanted it to stay that way. My colleagues agreed that love was a mystery, but they thought the senator should welcome all the help he could get in solving the mystery, given his own problems with divorce." (p. 96)

"Sociobiologists like E. O. Wilson believe that understanding the relationship between our animal qualities and our behavior frees us to improve our behavior, similar to the way that understanding the relationship between tubercule bacilli and disease freed us to find effective treatment for tuberculosis." (p. 210)

Ehrlich, Paul R. *Human Natures: Genes, Cultures, and the Human Prospect* (2000) *****
One of the best books on human evolution

This ambitious work, the magnum opus of Paul Ehrlich, Bing Professor of Population Studies and of Biological Sciences at Stanford University, and the popular author of the bestselling *The Population Bomb* (1968) and many other works, is quite a challenge for any reviewer. Perhaps I should begin with the footnotes. There are too many of them. On page 32, for example, Ehrlich writes "...twelve populations of a gut bacterium have now been tracked..." footnoting the word "bacterium." When one turns to the back of the book, one finds simply, "Escherichia coli." Perhaps it would be better to have written "Escherichia coli" in the first place! And something needs to be done about letting us know which footnotes are extensions of the text (afterthoughts, clarifications, etc., that we might want to go chasing after) and which are merely references. Of course this is a problem with all extensive works of scholarship. I suggest two sets of notes: one for references and one for clarifications, indicated perhaps by numbers for one and alphabetic letters for the other. There are 100 pages of footnotes here (1,909 altogether!) and 76 pages of works referenced. The index—a particularly good one, by the way—covers 18 double-columned pages.

Footnotes aside, this is a book that in a sense summarizes a long and much laureled career written by a man whose work and accomplishments we can all admire and respect regardless of whether we agree with his sometimes all too politically correct conclusions. Human Natures, despite Ehrlich's careful avoidable of such terminology, is about evolutionary psychology; that is, about how our understanding of evolution and our place as evolved and evolving beings affects how we view ourselves and our prospects. His "human natures" are derived from a survey of a vast literature including the work of anthropologists, ethnologists, sociologists, biologists, sociobiologists, psychologists, evolutionary psychologists, ecologists, demographers, geneticists, behavioral geneticists, historians, etc., etc. (This splintering and proliferation of disciplines as we enter the third millennium of the current era makes one long for consilience!) Most of these disciplines and many others have as their unifying principle the process of evolution. Ehrlich writes in the Preface (p. xi) "I want to show how a greater familiarity with evolution might contribute to our resolving...the human predicament." Ehrlich is referring to cultural evolution as well as biological. His primary thesis, that we are not of one "human nature" but of various "human natures," is implicit throughout. Made explicit on page 227, is his belief that since the beginning of agriculture 10,000 years ago we human beings have been, and continue to be, more subject to the forces of cultural evolution than we are to biological evolution. He writes, "cultural evolution...at this stage of our development begins to swamp the more gradual processes of biological evolution." It's clear that he wants to emphasize cultural evolution because we can do something about it, whereas to change our biological nature would involve human genetic engineering, a process that Ehrlich is understandably loathe to endorse (see especially page 66).

Ehrlich argues strongly for the plasticity of human behavior. He doesn't like generalizations about innate behavior. This can be seen particularly in his analysis of the causes of war in the Chapter 11. He insists with some exasperation that "It is senseless from any viewpoint for people to keep acting as if it were either possible or pertinent to determine whether human beings are innately aggressive or innately pacific" (p. 264). Ehrlich wants to encourage the idea that education can lead to more desirable behaviors. If our behavior is innate, then perhaps we can't be blamed for it, and furthermore (and worse) we can't do anything about it in terms of education, etc. It is this fatalism that Ehrlich is preaching against.

In short, Ehrlich takes what might be considered a caring, "liberal," politically correct view of human nature as opposed to others who see us merely acting out our genetically and culturally derived destinies. His position is comforting, but strange to say I really see little or no difference between his PC version of "human natures" and, e.g., Edward O. Wilson's much maligned amoral view. I think the differences are really matters of terminology and emphasis.

I want to add a couple of clarifications. Apparently Ehrlich is unaware of the full significance of "the handicap principle" advanced by Amotz and Avishag Zahavi in their book of the same name (1997) which in part explains altruism beyond kinship and reciprocity, namely that altruism is sometimes an advertisement to potential mates of one's fitness. Also in the chapter entitled "Why Men Rule" Ehrlich points to men being bigger as one of the reasons they rule, and to their being freer (because they have less of a reproductive burden) as another. But as Bobbi Low has pointed out in her *Why Sex Matters* (2000) men rule because there is no reproductive advantage to be gained by women in taking the reins of power. Whereas men have been able to gain greater sexual access to females by becoming rulers and thereby increase their reproductive fitness (Bill Clinton, notwithstanding), women gain little to nothing in securing access to more males.

Also there is the matter of the synchronization of menstrual cycles by women living together discussed on page 182. Ehrlich says the evolutionary significance is unknown. Actually it's fairly clear: in a harem situation if the females become fertile all at the same time, this provides an opportunity to more broadly mix the gene pool because the harem master can't possibly do it all himself, and so other males may get an opportunity. This embarrassment of riches for the harem master reminds me of the walnut tree producing so many seeds in a boon year that the squirrels can't possible eat them all.

This is a great book engagingly written by a man at the pinnacle of his career, a man I admire and respect.

Eldredge, Niles *Why We Do It: Rethinking Sex and the Selfish Gene* (2004) ***
An emotional response to evolutionary psychology

This is an attack on what Eldredge calls "ultra-Darwinism" and what he imagines is "selfish gene biology." The main problem in the first instance is that no such animal as "ultra-Darwinism" exists (it's just a slur); and in the second he is tilting against the windmill of a metaphor.

Richard Dawkins, celebrated author of *The Selfish Gene* (1976), is well aware that genes are not "selfish" in a literal sense. Furthermore, nearly everybody knows that genes work in concert with the environment to shape our biology and our behavior. Indeed, there is nary an evolutionary biologist outside of Bob Jones University who thinks that some kind of endowment, fixed or otherwise, is the exclusive determinate of who we are.

But Eldredge seems unaware of the modern understanding. Not only is he tilting at windmills, he is setting up and trying to knock down straw men that don't exist. Let's look at some of his accusations.

He wants us to know that the drive to eat and stay alive is more fundamental that the drive to reproduce. He calls this the primacy of economics over sex. This is fine, but I know of no evolutionary biologist, anthropologist or sociobiologist who thinks otherwise. They do not mistake the blueprint for the building. Of course in the mass culture a simplistic imbibing of Darwinism and a literal grokking of the metaphor of the selfish gene does exist. It is therefore perhaps a shame that some of this book does not appear in say *People* magazine to set the general public straight.

Eldredge notes that reproduction is NOT the purpose of life and posits the existential view that if life has a purpose "it is simply to live." (p. 46) But "purpose" is entirely an anthropomorphic notion and has no place in evolutionary biological thinking..

He wants to emphasize the cooperative nature of organisms as opposed to the idea that nature is competitive. He writes that "overt, no-holds-barred competition in the mating arena is, in the last analysis, relatively rare." And then on the very same page (66) he more or less contradicts himself by writing that male birds "stake out a territory (usually constantly defended against intruding males)..." Note that even using such ideas as "defended" ushers us into the land of metaphor. The birds actually react instinctively to the close proximity of other males and try to chase them away. *We* think they are "defending territory."

Eldredge is saying that the males are not fighting over females or sex but are holding onto valuable real estate—that is, their behavior is economic and not sexual. In a nut shell this is his point: life is lived primarily as an economic venture. What counts is getting enough to eat while avoiding life's many pitfalls. He believes it is a mistake to go further and add that the purpose of these behaviors is to reproduce. Again the bugaboo here is that word "purpose." The truth is that all organisms once they have secured the necessities of life try to reproduce. This is *not* the same thing as saying that is their "purpose."

What I especially dislike however is not Eldredge's insistence on what should be obvious, but the surly manner in which he attempts to dismiss certain of his colleagues and his attempt to ridicule ideas he either doesn't understand or thinks are being applied too broadly. His dismissive labeling—"hard-core evolutionary genetics," p. 130; E. O. Wilson's "consilience gambit" (why is it a "gambit"?) p. 249, "ultra-Darwinism," etc.—cannot stand for cogent argument. Particularly offensive is his repetition of what he calls "the Pleistocene cop-out." His argument here is that evolutionary psychologists explain current human behavior in terms of what worked on the savannas of Africa during the period of evolutionary adaptation. What he attempts to show is that our behaviors are culturally directed and not dances choreographed by genetic puppeteers. The truth is our behaviors are the product of both cultural and genetic influences working in concert.

Nonetheless our genetic heritage *is* in no small part the product of our experience during the Pleistocene, and it is part of the genius of evolutionary psychology to recognize this fact. Curiously Eldredge reveals that he understands this because on page 190 he writes (referring to Olduvai Gorge in East Africa), "It is the last best vestige of the environment that produced us, giving us insight into the very conditions in which our bodies—and behaviors—were shaped by evolution." That is pure evo psych, but apparently what Eldredge appreciates on one page is not always evident on another!

This is not to say that this book is without merit. Very well done is Eldredge's answer to the idea that rape is evolutionarily adaptive. (It is not: it is socially abhorrent for one thing; and since we are social animals, the rapist's behavior has met, and will continue to meet, with the severe disapprobation of society to the rapist's reproductive detriment. This is in addition to the fact that infanticide or neglect of the rapist's offspring is widely practiced.)

Also very much worth reading is Eldredge's exploration of infanticide and his explanation for its near universal practice throughout human history—although his point that it is not adaptive in an evolutionary sense is flawed. Sometimes it is better to have fewer children so that the ones we do have gain our full economic attention. The fact that this was often achieved through infanticide does not alter that general argument.

It is a shame that Eldredge's emotional need to discredit evolutionary psychology mars what could have been a useful exercise. He should have concentrated on arguments against rape as an evolutionary adaptation and eschewed the mistaken and gratuitous attacks on his colleagues. His general concern that those not expert in evolutionary biology sometimes overrate the genetic human endowment and underrate the cultural influence is a good one; but this point has to be made without straw men and ad hominem attacks, otherwise the author loses credibility and begins to sound more like a radio talk show host than a reputable scientist.

Note too that Eldredge's dedication is "In the spirit of Marvin Harris, 1927-2002." Harris was an anthropologist of unusual acumen who liked to debunk popular misconceptions, such as why cows are "worshipped" in India or why Muslims and Jews don't eat pork. He wrote such felicitous and readable prose that his books sold like novels. Well, Eldredge, to paraphrase a politician of recent memory, is no Marvin Harris. But he tries, and when he gets past his rather vague and somewhat mysterious confusion with the language of sociobiology and evolutionary psychology, and goes for the jugular, as it were, of the delusion of adaptive rape, he almost gets there.

Etcoff, Nancy *The Survival of the Prettiest: The Science of Beauty* (1999) *****
Readable, exciting, persuasive

The Survival of the Prettiest is an eminently readable, wisdom-filled, witty and very well-documented report on the human concept and experience of beauty and its utility, especially human beauty, or the perceived lack thereof. It is an example of a way of looking at ourselves that is becoming increasingly of value, both in terms of the insights it affords, and in the way it frees us from the muddled delusions of the past. This point of view is from the fledgling science of evolutionary psychology of which Professor Etcoff is a very persuasive spokesperson and practitioner.

"Pretty is as pretty does" and "Beauty is truth, truth beauty,–that is all/Ye know on earth, and all ye need to know" (Keats) are two widely differing attitudes toward beauty, but each in its way contains an essence of truth. However, rather than bring these or other presuppositions to what Etcoff has to say (as some readers have), I suggest we actually read what she has to say, and then draw our conclusions. What I predict will happen is that even the most ardent beauty-phobe will find something of value and enlightenment here.

Unfortunately (and understandably) not all readers have been able to approach the subject with an open mind. I noticed that an anonymous "reader" brought anorexia and bulimia into the discussion and blamed the rise in their instance on "media images" of beauty. No doubt media images are partly to blame (if indeed these disorders have become more prevalent). But it is more likely that the apparent rise in anorexia and bulimia is the result of the fact that the counseling professions now recognize that these eating disorders exist. In the past the symptoms had no commonly agreed upon locus such as "anorexia" or "bulimia" to adhere to, so we really do not know how prevalent they were. But more important in terms of being a public health problem is the enormous increase in obesity in this country, now often identified as an eating disorder due to "carbohydrate intolerance." The numbers of obese Americans hugely overwhelms the number of anorexics and bulimics, and obesity can hardly be blamed on "media images." We can point to the "supersizing" of fast food dispensers if we want to fix blame. However—and this is one of Etcoff's important points—it is not the media or advertising that is primarily responsible for our perceptions of beauty (or our tendency to eat too much), but an inborn, predisposition that has proven adaptive in the past that makes us find some people pretty and some others not so pretty.

Another "reader" claimed that Etcoff did not consider ideas of beauty in other cultures. That is incorrect, as anybody who has read the book knows. She devotes considerable ink to standards and ideals of beauty in cultures around the world and her observation is that ideals of beauty tend to be culture specific; that is, Ache tribesmen find their women and women of a neighboring tribe more attractive than European women. Indeed Etcoff reports that Asians typically find European and African noses not attractive because they are too large. Ache tribesmen actually made fun of the Caucasian anthropologists calling them "pyta puku, meaning longnose." (p. 139) Etcoff concluded that there were differences in standards of beauty, but that there were also similarities, and she goes into considerable depth detailing the studies. (See especially Chapter Five, "Feature Presentation.")

Etcoff is also criticized for her many literary quotes, references and allusions. But to my discernment they are a strength of the book and not a weakness. A very important part of our understanding of human nature comes not from the relatively new knowledge called science but from religion and literature. Etcoff is doubly wise to reference what great writers, statesmen and religious leaders have said about our ideas of beauty, first because what they say is worth knowing, and second because they express themselves so well. The anonymous reviewer who claimed to be a scientist perhaps ought to expand his or her reading to include wisdom from other sources, as has Etcoff. I just wish half of the writers writing today were one half as eloquent and readable as is Etcoff; and I'd settle for one-quarter as wise.

One of the significant things that this book does is to show that evolutionary psychology, despite the beliefs of its critics (and even that of some of its practitioners), is not limited to using insights from biological evolution alone, but from cultural evolution as well. Etcoff's book is a splendid example of this wiser, broader, synergistically more powerful employment.

Hamer, Dean *The God Gene: How Faith Is Hardwired into Our Genes* (2004)
****
Is "spirituality" an instinct?

Perhaps this book should be called "The Faith Gene" instead of "The God Gene." Geneticist Dean Hamer himself admits that "The God Gene is in fact a gross oversimplification...There are probably many different genes involved..." (p. 8) Later on he writes, "I believe our genetic predisposition for faith [notice: not "God"] is no accident. It provides us with a sense of purpose beyond ourselves and keeps us from being incapacitated by our dread of mortality." (p. 143) Note also that the book's subtitle declares that "Faith is Hardwired into Our Genes" while on page 211, Hamer declares that we are "Softwired for God."

Hamer's problem with definitions and usage arises because he is trying to take an abstraction such as "spirituality" or "transcendence" or "faith" or a belief in "God" and measure this abstraction with personality tests or by observing broader forms of human behavior. Furthermore he wants to make a useful distinction between religiousness and spirituality, between the extrinsic and intrinsic expression, the former being mostly public, such as church attendance, and the latter mostly private, such as prayer or meditation. Having done this he then wants to find a gene or some genes that code for spirituality. This is like trying to catch the ether in a hairnet.

Nonetheless it goes almost without saying that however ill-defined such abstractions may be, they do in fact refer to something real. A belief in an afterlife, in souls and inherited karma, in gods and poltergeists, heavens and hells, in things mystical and extrasensory, in a reality beyond a purely material and animal existence is universal to all human societies, past and present, and would seem to be as necessary as the very air we breathe. (Gurus, churches and religions exploit this human necessity.)

Consequently it is not so far-fetched to look for the predisposition for such beliefs in our genetic code, genes that have been selected by the evolutionary process. The question remains however, exactly what behavior is it that is selected and found adaptive in an evolutionary sense? Hamer thinks it is some sort of personal transcendence—that is, spirituality as opposed to religion as such (see page 215). However I think there is reason to believe that what is selected is the more profane aspects of religion and spirituality. To put it bluntly, what the genes (interacting with the environment of course) code for are tribalisms such as following a leader and being willing to die for the good of the tribe, and in general following the authority of tribal ways and means, believing what the shaman says, what the priest says, what the ayatollah tells us, and what the documents of the tribe declare as true.

Edward O. Wilson in his book, *On Human Nature* (1978)—highly recommended, by the way—argued that the ability of the individual to conform to the group dynamics of religion was in itself adaptive. He added, "When the gods are served, the Darwinian fitness of the members of the tribe is the ultimate if unrecognized beneficiary." (*opus cited*, p. 184)

Still there is a sense in which it is possible to see the genetic predisposition toward faith and religion in a more morally positive sense. Hamer believes that "God genes...provide human beings with an innate sense of optimism." (p. 12) Clearly life must be worth living, and faith provides us with any number of compensations for a hard life: promise of an afterlife, a rebirth to a better station, karmic comeuppance for transgressors, and karmic reward for our perceived good behavior, punishment for sin, etc., are artifacts of faith and are the main tenets of many religions.

However as to the specific gene that Hamer comes to identify, the VMAT2 gene, which influences the flow of monoamines in the brain, it could be said that this gene is not so much the "God gene" as the "dope gene," the gene that helps us to get high. On page 77 he allows that "There might be another 50 genes or more of similar strength."

In addition to Hamer's central argument, there are aspects of this book that are interesting and valuable in themselves. The chapter on "The DNA of the Jews" is absolutely fascinating and gives us a good idea of what is possible by using the changes in either the "y" or "x" chromosomes to trace human migrations and intermarriages.

I also like the distinction that Hamer makes between spirituality and religion. We all know people who are spiritual, but don't go to church (or temple or mosque, etc.). And we all know people who attend church regularly but are about as spiritual as hyenas. (I won't mention any White House occupants, past or present!) And it is clear that there are agnostic scientists who are very spiritual persons indeed.

However, the weakest part of the book involves Hamer's attempt to adequately define spirituality and to distinguish it from religion. He calls in the psychology and psychiatric establishments to help out. I don't think they help much. It is a daunting task to even define "God" adequately. In the final analysis he goes with the idea of transcendence. However what we humans want to transcend is our animal nature (and sometimes the evidence of our senses and our experience!). Part of the reason we wear clothes and otherwise cover up while imagining that we have souls and are made in the image of God is to make our animal nature less obvious. For human beings it is not sufficient to be just animals. We are (or should be) spirit as well. Hamer actually declares that "Spirituality...is, in fact, an instinct." (p. 6)

Finally, faith does not require a god. Taoism has "the way," and the Buddha famously turned aside questions about God as being beside the point, while the ineffable God of the Vedas is nothing like a personal god.

Lawrence, Paul R. and **Nitin Nohria** *Driven: How Human Nature Shapes Our Choices* (2002) *******
Adam Smith discovers evolutionary psychology

This is evolutionary psychology as seen by two professors from the Harvard Business School (!). While some readers may be familiar with a lot of what is presented here, it is agreeable to get a perspective from another academic discipline and a new sense of application. It is especially pleasing because professors Lawrence and Nohria write well and have an appreciation of what an exciting time of biological discovery we are living in, a time when the convergence of knowledge and techniques from various disciplines is giving us the ability to look inside the black box of human nature previously closed.

The authors' use of the term "drives" to designate the source of behaviors is familiar, but the idea that these drives come from modules in the brain, or a network of modules, is what is relatively new. Whether this is just another construct like Freud's ego, id, and superego is an open question. However—and this is important and at the very essence of what is going on in brain science today—unlike Freud's construct, the one presented here is based on something tangible in the brain's structure. As the authors report, recent advances in technology allow us to discern the brain's structure as it works. These observations provide a scientific basis for constructs attempting to explain human behavior. Whether there are four fundamental drives, as Messrs. Lawrence and Nohria think, or some other number, or whether an entirely different construct is required, is also an open question. Personally, I find their array persuasive, and I think the idea of "drives" a valuable one. More important though is their understanding that we are motivated by more than rational self-interest, the so-called "invisible hand" from Adam Smith and the market place.

Here are the drives as defined on page 10:

D1 is to acquire objects and experiences that improve our status relative to others.

D2 is to bond with others in mutually beneficial, long-term relationships.

D3 is to learn about and make sense of ourselves and the world around us.

D4 is to defend ourselves, our loved ones, our beliefs, and our resources.

In should be noted that these four drives do not in any way contradict the general finding in biology that individuals tend to behave in such a way as to enhance their reproductive success. What is new is that such "selfish" behaviors include behaviors usually seen as altruistic. Yet I think the authors would enhance their understanding of the idea of "altruistic behavior" by reading Amotz and Avishag Zahavi's *The Handicap Principle: A Missing Piece of Darwin's Puzzle* (1997) in which the adaptive function of some altruistic behavior is to directly advertise fitness.

It should also be noted, as the authors do on page 63, that "What drives behavior is a contest among the emotions, not the rational calculation alone." In other words, rationality leads to the creation of an emotion which competes with the instinctive emotion. This is an important concept. It is not the rational mind overcoming the emotional mind, but the employment of emotion by the rational mind to overcome instinctive imperatives which sometimes lead us in the wrong direction.

Through the process of "social bonding" as presented on page 83, the authors embrace the idea of group selection, an idea disparaged by notions from Dawkins's "selfish gene" and elsewhere. The idea that there could be the selection of genes that "orient behavior toward the good of the group" has long been discounted by the establishment in evolutionary biology. (This view is changing.) The seemingly very convincing argument has been that "any carrier with a genetic disposition to be nice to others would be, in time, wiped out by the selfish free-riders in the population." (Still on page 83.) My feeling, however (similar to that of the authors), is that for human beings the "in time" part has never had a chance to kick in. This is mainly because of the constant struggle of tribe against tribe throughout human and pre-human history. The benefit to the tribe from individuals willing to risk life and limb for the good of the tribe is clear. What has not been realized by many is that the benefits to the individual by enhancing the tribe's fitness more than offset the loss incurred from taking risks. True, if the tribe faced no outward danger for a long period of time, the genes of the "selfish free-riders" would predominate in the population and the altruistic genes would die out. But that hasn't happened. Consequently groups (bands and tribes) that contained many "altruistic" individuals survived while groups with fewer altruistic individuals died out. Therefore we have the "group selection of individuals" (which is a way I have seen this phenomenon phrased).

I should also like to note that religion, the cultural evolution of, is accounted for in a similar way. Those tribes that had religious beliefs strong enough to facilitate bonding and altruistic behavior survived more often than tribes that did not. This is something that Edward O. Wilson pointed out some years ago in his book *On Human Nature*.

I think this is an excellent book for the general reader and a fine melding of the ideas of evolutionary biology into the culture of the work place and other loci in the modern world. The authors do a good job of showing how the ideas of evolutionary psychology go far beyond the retelling of "just so" stories, ideas that can help us to understand ourselves and the world in which we live.

Lewontin, Richard *It Ain't Necessary So: The Dream of the Human Genome and Other Illusions* (2000) ****
Maybe it is

This book is a collection of nine essays from The New York Review of Books, beginning in 1981, mostly on genetics, the genome and the Darwinian pantheon. The essays are presented with new footnotes and cross references followed by an Exchange and/or an Epilogue in which the material is updated and some contrary points of view presented and addressed. The expression is erudite, polished and complex, the tone authoritative and at times slyly satirical and at least a microbe's breath away from the pompous.

The first essay, "The Inferiority Complex" is a review of Stephen Jay Gould's *The Mismeasure of Man* (1981) which deals with the IQ conundrum. Lewontin's main point here, in agreement with Gould, is that "there may be genes for the shape of our heads, [but] there cannot be any for the shape of our ideas" (p. 9). I'm not sure I agree with that rousing call to the uniqueness of human kind, but I am confident that no one has yet refuted such a point of view. Not entirely as a surprise Gould (in a jacket blurb) acknowledges

Lewontin as "the smartest man I have ever met."

Gould is not the only one to sing praises to Lewontin's intellect and under-standing. Noam Chomsky chimes in with an acknowledgment of "the impres-sive quality and significance" of Lewontin's essays, while a book I just finished reading, Steve Jones's excellent *Darwin's Ghost* (1999) is dedicated to Lewontin, who showed him "what evolution can and cannot explain." Perhaps that is Lewontin's main strength, as an anchor on the ship of biological pre-sumption that would sail us to a questionable nirvana of the pre-determined. I can say from my own experience that the very learned professor reminds me of someone I would call "the Edmund Wilson of book critics biological." He is also the very distinguished Alexander Agassiz Research Professor at Harvard and the author of several books on genetics and related subjects, most charac-teristically perhaps, *Not in Our Genes* (1984) with Steven Rose and Leon J. Ka-min.

Why then am I not entirely thrilled with this beautifully wrought collection of unquestionably significant and stimulating essays? I think it's that I disagree with his point of view and emphasis, and feel that the sequencing of the hu-man genome really is a significant step toward our understanding of who and what we are, and I don't care who, or who did not, get rich in advancing it. I also think that the practical applications from such information may prove valuable in ways we cannot begin to predict. I am a fool for knowledge if only for knowledge's sake, and I wonder why Lewontin has expended so much en-ergy knocking the project. His real criticism of the effort, despite his use of the derogatory words, "dream" and "illusion" and even "fetish" (p. 135) is pre-sented on page 177: "The promise of great advances in medicine, not to speak of our knowledge of what it is to be human, is yet to be realized from sequenc-ing the human genome."

Who could disagree with that? He also writes on page 151, "Causal stories are lacking...nor is it clear, when actual cases are considered, how therapies will flow from a knowledge of DNA sequences." Again, who could disagree? How-ever this is political-speak. It says nothing that can be seized upon and found derogatory, yet hints at failure and disappointment. Characteristically, Lewontin writes nothing that one can find direct fault with, yet by indirection and association he belittles the effort. I would note that the word "fetish" is not used directly as a coloration of the project, but as an indirect association.

People have said that The New York Review of Books is really The New York Review of Each Other's Books, and therefore constitutes a close-knit club with a shared political point of view. I will withhold such a judgment since I have only a passing familiarity with that prestigious publication.

Putting all that aside, I found myself, while reading the third chapter, "Darwin, Mendel, and the Mind," wondering if Lewontin was really conscious of his own thought processes when on page 103 he relates that he "passed among three very different mental states all under the control of the willful I." Ah, if only that "willful I" really was in control and had the power to consciously regulate our mental states. Lewontin seems unaware that it takes many years of devoted practice to still the "monkey mind" and allow one an observation of one's mental processes. He asks rhetorically (still on p. 103), the question he calls the "central problem...for neurobiology," namely, "What is "I"? This is indeed a profound question, asked at least as early as the Upanishads. The modern answer, which Lewontin must know, but does not present, is that the I is an illusion that we cannot help but believe. He goes on to argue with Daniel Dennett against the idea of consciousness as a "metaphorical delusion" (p. 105) without once realizing that there is a crucial difference between a "delusion," metaphorical or otherwise, and an illusion. If he looks more closely he might find that consciousness is a trick of the evolutionary process, the main purpose of which is to make us fear death by forcing us to identify intensely with our particular phenotype. Our subjective appreciation of consciousness is a wondrous byproduct of that identification.

Low, Bobbi S. *Why Sex Matters: A Darwinian Look at Human Behavior* (2000)
*****

Extraordinarily thorough, authoritative, and current

This book is not as formidable a reading challenge as might be supposed on first perusal. True it is 412 pages long, but the back matter begins with the footnotes on page 258. There follows a glossary, a 57-page bibliography, an author index and a subject index. Also, even though this is clearly an academic tome written by a professional ecologist who is not about to compromise her standing in the scientific community for a shot at popular success, Professor Low nonetheless employs a readable and common sense approach with a minimum of unnecessary jargon. Furthermore, what she has to say is exciting and relevant to our lives, and we can see that she cares about communicating to the reader as much as pleasing colleagues. Reading *Why Sex Matters* is consequently one very engaging experience.

Low, who is a professor of ecology at the University of Michigan, assumes the point of view of an evolutionary biologist as she asks the question, how are men and women different and why? She is particularly focused on how the sexes differentially use resources to further reproduction, and asks which behaviors are ephemeral, due to present conditions, and which are more enduring, having proven adaptive over longer periods of time and in differing environments. She faces squarely the unsettling feeling that some people get when they contemplate humans purely as biological entities—or "critters," to use her expression. As she tells us in the preface, there are three themes guiding her work: One, "resources are useful in...survival and reproduction"; two, "the sexes...differ in how they...use resources"; and three, "each sex accomplishes these ends" by reacting to the environment differently. The result of this structured approach is a clear introductory course in sexuality from an evolutionary point of view, and a fascinating read.

Because Low employs resources from a wide variety of disciplines, including sociobiology, evolutionary psychology, behavioral genetics, ecology, anthropology, sociology, biology, history, etc., not to mention pop culture and world literature, her work is highly persuasive in a scientific sense. And because she studiously avoids squabbling among the disciplines, her work is psychologically compelling. There is material on cultural transmissions as well as natural selection. Demographers are given currency along with those of evolutionary biologists. One gets the sense that she has read just about everything and has thoroughly evaluated what she has read. Particular interesting to me is her discussion of the tangled origins of sexuality and the (non-obvious) nature of altruism. The chapters on warfare, "Sex, Resources, and Early Warfare," and "The Ecology of Warfare" are worth the price of the book alone. There we see that women warriors are rare because men can gain reproductive advantage through warfare but women cannot (p. 216). Low suggests that war may be an example of "runaway sexual selection" and its practitioners may have become "unhooked" from the old reproductive rewards, but that the proximate rewards remain. Low soberly faces the prospect of future warfare when small groups of people may acquire monstrous weapons, noting that "given a short-term gain...versus an unspecifiable risk of nuclear warfare...in the future, we do not predict restraint."

It should be clear that Low is a professional academician and not a journalist as some popular writers on evolution are (Matt Ridley and Robert Wright, to name two of the best), and as such careful about her assertions. She doesn't espouse pet theories that may be overturned tomorrow; but she isn't afraid to voice her opinion. To give you a sense of her careful style, note the stunning qualification in the parenthetical in this statement from page 217 (and the sly irony): "Human war can become more complex and varied than intergroup aggression in other species, largely as a result of the development of technology (which itself is probably a product of intelligence)." Probably, indeed.

In the chapter on "Politics and Reproduction" we learn that men seek political power for reproductive gain (p. 211) but in the modern nation state may have to settle for proximate gains (which may be an irony not lost on Bill Clinton). Women, however, can gain little or no reproductive advantage directly for themselves, which may be the reason there are relatively few women in positions of political power in most human societies.

Some of this I admit is tough going. The material on "The Group Selection Muddle" in chapter nine is still muddled in my mind, and I couldn't figure out the point of the Summary of Selection Theories (Table 9.1 on pages 156-157). But evolution and the disciplines that address human nature are complex, in some ways, deceptively so.

Professor Low is wise, temperate, thorough and more objective than seems possible in such a vibrant and contentious academic field. I suspect that this book started out as an undergraduate text, but somewhere along the line those reading the manuscript realized that it was so interesting and valuable that it could be published as a trade book aimed at a general readership. If you have time to read only one book on human nature, read this one. You will learn more than you would from half a dozen "popular" expositions, and you will have a sense of having learned something important and valuable. I wish I had known what is in this book when I was one and twenty. I would have conducted my life with a lot more grace and effectiveness.

Miller, Alan S. and **Satoshi Kanazawa** *Why Beautiful People Have More Daughters* (2007) ****
"Women are the reason men do everything" (p. 133)

Many years ago, before evo psych was even sociobiology, some people (usually social scientists) would ask themselves, how did things go down in the prehistory? They realized that our instinctive behaviors were honed on the savannahs of Africa long before we became civilized or even before we became human. The Darwinians among them further realized that the ten thousand or so years since the beginning of agriculture and animal husbandry was not enough time for human nature to have changed much. Ergo, we are savannah animals dining at the Burger King with our fingers on the nuclear trigger reading the Wall Street Journal, but with our biological imperatives virtually unchanged since the Stone Age.

From that simple, but profound, realization has sprung evolutionary psychology, which is a fine tool for gazing more or less objectively into the labyrinth of human behavior leading to some understanding of why we behave the way we do.

As wonderful as I think evolutionary psychology is—and it is indeed an eye opener that has taken the groves of academy by storm in the last couple of decades—I can readily see five problems:

One, it upsets people much in the same way that Freud or Darwin upset people, namely by making us more like animals than like beings made in the image of God.

Two, evolutionary psychology, like all psychologies, is limited.

Three, sometimes it is difficult to tell the difference between something obviously true (men want lots and lots of reproductive opportunities) and something that may be true ("the death penalty cannot deter young men" from violent crimes—see page 130).

Four, the unwarranted leap that many people, even some very intelligent and educated people, make from the *is* of an evo psych discovery to the *ought* of a moral or societal truth; e.g., women want a man committed to helping them raise their children, but they also want the genetic input from the most alpha male they can find. This, to many people, makes it sound like cuckolding your hubby is the right thing to do since it is the "natural" thing to do. It is also the natural thing to take what you want when you want it, but that doesn't make it right.

Five, behavioral tendencies as gleaned from a study of humans in the so-called Environment of Evolutionary Adaptation are just that, general tendencies that most people at one time or another, for a myriad of reasons, do not always follow. Evolutionary psychology describes main tendencies; it does not prescribe anything. Of course some of these tendencies are powerful biological imperatives that most people find difficult to ignore.

The strength of this book is that the authors go well beyond the familiar discoveries from evolutionary biology to lesser known but fascinating discoveries such as the evolutionary rationale behind beautiful people being more likely to have daughters than sons, to why rich people are more likely to have sons, or why having sons reduces the chance of a divorce, to even why gentlemen prefer blondes.

Here are some observations on the few cases I think the authors didn't get quite right:

They ask: "What is the adaptive problem that religion is designed to solve? Do religious people live longer or have greater reproductive success? So far, no one has been able to point to an adaptive problem that religion is designed to solve." (pp. 158-159) Not so. As Edward O. Wilson so eloquently put it in *On Human Nature* (1978): "When the gods are served, the Darwinian fitness of the members of the tribe is the ultimate if unrecognized beneficiary." (p. 184) What he meant was that the adaptive reason for religion is to make the tribe more cohesive and better able to defeat other tribes in, for example, warfare.

The authors write: "The reason most Western industrial societies are monogamous, despite the fact that humans are naturally polygynous, is that men in such societies tend to be more or less equal in their resources, compared to their ancestors in medieval times." (p. 90) While I suppose this is true, a better reason is that large polygynous societies are politically unstable since large numbers of males without mates tend to revolution; and given suffrage, they would vote against polygyny, as in the US.

The authors aver that there is no satisfactory (adaptive) explanation for why soldiers die for their country. (p. 186) The clear explanation is that young men put themselves in positions in which they are likely to die in battle because society sees that as being brave and manly, and females like to mate with brave and manly men. The fact that many of these men might die before re-producing is offset by the increased reproductive fitness of those who don't die and the fact that they often (as the authors report) have sex before going off to war.

Another bugaboo that authors don't believe is answered is how homosexuality can be adaptive. (See page 180.) The simple answer is that homosexuality in many environments leads to effective male bonding which in turn can lead to a monopolizing of the available females. While homosexual men may not cop-ulate with the females as much as their heterosexual buddies, they will none-theless copulate a lot more often than loners who do not have access to the females.

One more point: many sociologists might object to the authors' use of the term "Standard Social Science Model." Not being a sociologist myself, I find it hard to believe that the Standard Social Science Model, as characterized by the authors, virtually ignores evolutionary biology and sees everything in pure-ly cultural terms, leaving us to believe, for example, that gender differences in male-female behaviors are largely the result of a patriarchal bias in society.

Written in a popular style with some understandable simplicity, this book is an excellent introduction to evolutionary psychology, nee sociobiology, which, along with cognitive psychology and neuroscience, constitutes the essence of contemporary academic psychology.

Miller, Geoffrey F. *The Mating Mind: How Sexual Choice Shaped the Evolution of Human Nature* (2000) ****
Sexual selection writ large, very large

There was a commercial for Toyota's Paseo a few years ago in which its reliability was being touted when suddenly an insistent voice came on and whispered what Toyota hoped was the subliminal truth: *"Women dig it!"*

I think the perception of the human brain as a sexual ornament—Geoffrey Miller's primary argument in this book—is in the same class. Sex may help to sell cars, and the fact that women are attracted to men of means with expensive cars (forget the Paseo) is not to be doubted, but to suppose that the primary function of any automobile is symbolic or ornamental is mistaken. While witty conversation and musical display certainly are attractive attributes of a well-made brain, they are secondary to the social, political and subsistence skills of that brain as reasons for its being.

The brain, as Miller points out, is a very expensive organ, eating up a disproportional amount of our caloric intake, requiring a long period of development both inside and outside the womb, as well as making childbirth painful and dangerous—which are some of the reasons it has never grown so disproportionately large in any other creature. Such an organ must have some very fine compensating qualities to make it adaptive. The fact that women dig it (or men for that matter) isn't enough. The case of the peacock with its huge showy tail feathers and the Irish elk with its enormous antlers are cited as examples of runaway sexual ornamentation, and they are that, I suppose. But the Irish elk is extinct and the peacock is not exactly a favorite to survive for very much longer. Most creatures do not develop features that are so grossly expensive strictly as lures for the opposite sex. Most such lures are modest, the coloring of birds, the combs on roosters, the mane on a lion, a woman's breasts, etc.

Miller knows this and his book does not contradict what I have written. The problem is one of perception and understanding. Evolutionary theory, seemingly so simple at first glance, is an incredibly complex subject, so much so that no one person can hope to grasp it all—or, I should say, grasp what little we now know. Evolutionary psychology, which follows from the fact of evolution just as surely as evolutionary biology does, is also an incredibly complex subject awaiting its first genius. But separating "just so stories" from genuine insights is extraordinarily difficult. There is no way Miller or others can prove that the human brain developed in part because of sexual selection—although I personally do not doubt that it did. Nor do I doubt that the elephant's size

119

was also enhanced through sexual selection, perhaps even the cheetah's speed as well. But how to prove it? Consequently, it becomes a matter of opinion to what extent the thesis is correct. Is Miller overstating the case? There is no sure way of deciding. Each reader must weigh the arguments and reach his own conclusions. This is why evolutionary psychology, despite its enormous power to provide insight into who we are and why we behave as we do, is difficult for some people to understand and will continue to be controversial for decades to come.

But this very readable and thought-provoking work goes well beyond its primary thesis as Miller explores sexual selection theory over a wide range of human ability and ornamentation. One of the most exciting things he does is to recognize that biologist Amotz Zahavi's handicap principle can function as a fitness indicator not only to predators but to members of the opposite sex. Thus springboks jump into the air to show predators that they have energy to spare, while women incur the "handicap" of growing large, fatty breasts to show that they have foraged well and have reserves to feed their children. In this way fitness is demonstrated, the former to discourage a fruitless chase, the latter to advertise ready fertility to members of the opposite sex.

I also appreciate his recall of neotony—the retention of child-like features in the adult—as a significant feature of sexual selection. Most recent books on evolutionary psychology have forgotten Ashley Montagu's *Growing Young* (1989) and Stephen J. Gould's *Ontongeny and Phylogeny* (1977) both cited by Miller. Sexual selection makes us appear more childlike as a youth indicator, consistent with choosing a mate that will serve us well as a partner for many years to come.

Miller will find his critics here because one of the things he is trying to do is diminish the idea that the intellectual and artistic abilities of the brain are merely fortuitous side effects of its subsistent and social function, which is the position of Stephen Jay Gould and others. In this attempt I don't think Miller is going to be entirely successful. Fortuitous side effects of organs are at the very heart of evolutionary change, sometimes becoming more important than the earlier function. We see this in the case of fins that became legs, or light-detecting organs that became capable of discerning movement and color, and so on. The best explanation for the rapid and disproportionate growth of the human brain is its social/political function. Following in importance is its ability to understand the environment in a way that allows us to find food and shelter and avoid predators. Then, I would say, its tertiary function is to display itself to the opposite sex, to be "sexy."

It is impossible to do justice to this ambitious work in a review limited to a thousand words; however had I more space, given the wide terrain explored by Miller, I would be inadequate to the task. Quite simply, this is an incredibly complex work by a gifted young scientist trying to establish himself as one of the leaders in the growing field of evolutionary psychology, and a book all interested persons should read.

Ridley, Matt *The Origins of Virtue: Human Instincts and the Evolution of Cooperation* (1996) ****
Lively, biased, and a whole lot of fun

Matt Ridley nicely demonstrates here that there is no such thing as virtue and that altruism is an oxymoron. Instead it is all reciprocity and enlightened self-interest. This reminds me of when I was a sophomore in college. We used to argue passionately about three things: the nature of women, whether the Pope believed in God, and whether it was possible to act otherwise than in one's own self-interest. We concluded that women were an enigma wrapped in a mystery, etc.; that it wasn't clear whether the Pope believed in God or not; and that, barring mistakes, we always acted in our own self-interest. We further concluded that "altruism" was a word without real meaning, that the Pope was an amoral political animal, and that women were, regardless of their nature, *very* interesting. But we were sophomores. Matt Ridley is all grown up, and what interests him in this book is not so much the origin of virtue (although he does get heavily into that) but the restoration of the conservative agenda. Alas. He argues from biology (our nature) to what ought to be politically. This is doubly "alas" because Ridley preaches mightily against this very delusion, calling it a "reverse naturalistic fallacy" (p. 257).

David Ricardo and Adam Smith are brought into the fray, Hobbes and Machiavelli. Ridley takes arguments from game theory and political science and the world of high finance to make his point that virtue as it is ordinarily understood does not exist. He goes on to call for less government and more local autonomy, a return to a dream state of "everything small and local" (p. 264).

As he does, Ridley comes dangerously close to taking on all the trappings of a right wing radio talk show host, spouting the virtues of Newt Gingrich and Margaret Thatcher on his way to becoming something like a high-toned Rush Limbaugh.

Alas, how sharp was his rapier and how telling his prose when Ridley stuck to revealing our social and sexual hypocrisy as he did so delightfully in *The Red Queen* (1993); but how obvious are his prejudices when he steps into the political arena. He actually argues that tried old irrelevancy of the embarrassed right wing, that even though Hitler was bad, very bad, he was better than Stalin. Thus on page 258 we have (referring to the doctrine of acquired characteristics embraced by the Soviet state): "Unlike the genetic determinism of Hitler, Stalin's environmental variety went on to infect other peoples."

Ridley even argues that Hitler got his ideas from the communists. "Hitler was merely carrying out a genocidal policy against 'inferior', incurable or reactionary tribes that Karl Marx and Friedrich Engels had advocated..." (p. 253). So caught up in his cause is Ridley that he begins to contradict himself and argue for the kind of idyllic fantasy world that he condemns in Rousseauians. Thus in his chapter entitled "The Power of Property" he waxes nostalgic for the "egalitarian" conservation systems of New Guinea fishermen and Maine lobster men before the interference of big government. On page 262 he talks about "The collapse of community spirit in the last few decades, and the erosion of civic virtue...caused" by "the dead hand of the Leviathan." But on the very next page he declares, "I hold to no foggy nostalgia that the past was any better. Most of the past was a time of authority, too..."

Yes, Matt, it was. The authority of the gang lords, of the feudal lord, of a system of social, political and economic imprisonment so oppressive that the average person never got further than a few miles from the place of his birth and had little to no chance of rising above the economic and social station of his birth. It was "small and local" with a vengeance. The tyranny of the feudal lords in Europe and, e.g., the war lords in China is conveniently ignored in Ridley's political fantasy. He claims that we have it better today only because of superior technology (p. 263) forgetting that our system of representative democracy in Republican form is also an improvement over the absolutism of the tribe. The sad lesson here is, that even a man as adroitly talented and as intelligent as Matt Ridley becomes just another propagandist when he ventures into an area in which he is emotionally involved.

Still there is a lot to enjoy in *The Origins of Virtue*. His discussion of the prisoner's dilemma is the best I've read, although his analysis of the "wolf's dilemma" (p. 55) is faulty. I won't go into it here, but "the tiny chance" that he refers to is overwhelmed by the fact that each player has only a five percent chance of "winning" by pushing his button since he has to beat 19 others to the punch. Consequently the best strategy is the obvious, don't push that button! (But check this out for yourself.) His discussion of how the division of labor has enriched our world is interesting; his analysis of how we detect cheaters and how that is an instinctive human talent is persuasive; and his delineation of the nature of gift giving and receiving and how it relates to our innate sense of reciprocity is valuable as it shines light on the nature of "virtue." In fact, his entire argument is eminently worth reading. His glorification of trade (with which I agree) and his put down of ecologists (with which I disagree) is tolerable. Most fun though—recalling the Matt Ridley of *The Red Queen*—is in all the sacred cows he slaughters along the way: the New World Indians (ouch!), Margaret Mead, the so-called "tragedy of the commons" theory, the Noble Savage, even poor Chief Seattle is revealed as a slave-owner whose public reputation is largely the product of a screenwriter's imagination (p. 214).

Ridley, Matt *The Red Queen: Sex and the Evolution of Human Nature* (1993)
*****

Informative, witty and fun to read

This is the book that first demonstrated to me the power of evolutionary psychology to help us understand ourselves. Published a year before Robert Wright's *The Moral Animal*, which covers much of the same territory, this is to my mind a more sophisticated and more direct exposition. Both books are characterized by a sly wit and an incisive expression, but Ridley meanders less among the relics of Freud and Darwin and is less concerned about whether we're moral or not and more concerned with what's sexy and why. He had a lot of fun with this book and it shows.

The "red queen" is a metaphor for an arms race. In an arms race both sides run as fast and as hard as they can to stay in the same place relatively speaking. In evolution the arms race is between parasite and host or between predator and prey. Both are running as fast as they can just to keep up, because when one gets an advantage, the other finds a counter. The red queen comes from Lewis Carroll's *Through the Looking Glass and What Alice Found There* (1871) since that monarch ran as fast as she could but never got anywhere at all. The red queen is also a metaphor for the theory that there is no "progress" in evolution, that "...species do not get better at surviving... Their chances of extinction are random." (p. 64)

Ridley covers a lot of territory here, ranging from sex to the handicap principle to gossip to why our brains are big (to figure out what the other person is up to!). *The Red Queen* answers the question, "Why is there sex?" Apparently we have sexuality rather than asexuality because of the arms race between microbes and our immune systems. Sex is a way of storing defenses against parasites in the gene pool of the species and then mixing them anew each generation to fool the microbes. Without the gene pool and the DNA mixing, the microbes would quickly evolve a way around the organism's defenses; but with sexuality the organism juggles its "locks" every generation and so is able to keep up with the fast-mutating microbes. When again the microbes evolve the keys to these locks, the gene pool is mixed again and the organism comes up with an old lock that the microbes again have to evolve a key to.

Some of the fun is the incisive way Ridley presents the ideas, and the ideas he chooses to present. For example, note how effectively he demolishes Freud's naive incest taboo theory on pages 282-286. Also interesting is his presentation of the idea that it is not thinness in women per se that attracts men, but a low ratio of waistline to hip line that fetches them. There are chapters entitled "Polygamy and the Nature of Men," and "Monogamy and the Nature of Women." In Chapter 9, "The Uses of Beauty," Ridley goes into some detail on why men prefer thin and blond women. And on pages 217-218 he explains why women cuckold their mates: "This is because her husband is, almost by definition, usually not the best male there is—else how would he have ended up married to her?" She wants the parental care of her husband and some other man's superior—she thinks—genes.

Ridley is rather modest and says that most of the ideas in the book are not his and at any rate many of them will undoubtedly be proven wrong. This is refreshing to read when I think about all the delusive ideas so proudly trumpeted by popular books on evolution and human behavior in the past. Desmond Morris's *The Naked Ape* (1967) and Elaine Morgan's *The Aquatic Ape: A Theory of Human Evolution* (1982) come to mind, both fine books, but now seen to be substantially mistaken.

Written in an engaging and lucid style, *The Red Queen* really is the best of a number of books on evolutionary psychology to appear over the last decade and one that is a delight to read.

Rose, Hilary and **Steven Rose**, eds. *Alas, Poor Darwin: Arguments against Evolutionary Psychology* (2000) **
Alas, Poor Roses

One is hardly onto to the second page of this misguided, miss-conceived and miss-edited book than the illogic and misrepresentations begin. The authors begin by branding evolutionary psychology "henceforward EP" as "a particularly Anglo-American phenomenon," and reference this claim with a footnote on page 16 stating that "Other European countries, notably France, have been less overwhelmed by Darwinian evolutionary theory." One wonders at the point of this. How does it play in China or Japan? On the other hand, maybe they're after evolutionary theory itself and not just EP! One also wonders what the acceptance of "Darwinian evolutionary theory" by some countries and not by others (even if that was somehow demonstrated) has to do with a critique of evolutionary psychology. If France has not been "overwhelmed by evolutionary theory" perhaps the worst for France. And who says Anglo-Americans have been overwhelmed by Darwinian evolutionary theory? Most Americans, at any rate, still believe in angels!

This sort of slippery, non sequitur-filled prose is, alas, typical of much of what follows. Here's another quick example, in a footnote on page 16. The authors reference a poll from the journal *Science* showing "that a great majority of life scientists are now non-believers." Incredibly, they follow this immediately with the non-logical: "Physicists are less hostile to religion..." seemingly innocent of the fact that being a non-believer does *not* necessarily imply hostility to religion! Still on page two the authors write, "It [evolutionary psychology] claims to explain all aspects of human behavior...on the basis of ...features" formed "during the infancy of our species some 100,000-600,000 years ago."

a very fine creative writer. But I would remind him that *all* psychologies are limited, and so to criticize evolutionary psychology for being limited in what it can tell us about ourselves, isn't much of a criticism, unless one can also argue that cognitive, or behaviorism, or psychoanalytic, or some other psychology is superior. I happen to think, as I have said before, that the psychologies found in the great world religions are superior to any of the academic psychologies. Only evolutionary psychology is able to offer something with the same kind of antediluvian power. Jencks does not mention other psychologies. He does claim that Wilson, at the end of his lecture, went from the "is" of evolutionary psychology (which is really how to understand and appreciate EP) to the "ought" of a moral imperative, the classic error incidentally made by many who criticize EP. Problem is, Jencks doesn't reveal what Wilson said that convinced him that Wilson had gone too far (maybe I missed it). I do know from having read several of Wilson's books that he knows better than to fall into that trap. I wish I could say the same for the Roses.

This is by way of signing off: I am about to exceed Amazon.com's 1,000-word limit. Some of the other reviews at Amazon comment very intelligently on some of the other essays. I particularly recommend the reviews by Todd I. Stark and A.P. Jackson.

Rose, Steven *Lifelines: Biology beyond Determinism* (1997) ***
Argues against reductionism in biology

We are not objects. We cannot be defined by our genes. It is only through an understanding of our developmental history in interaction with our environment that we can hope to know who we are.

Thus Steven Rose, molecular brain biologist and staunch foe of reductionist biology, has called upon the metaphor "lifelines" to describe our "trajectory" through time and space. We are processes. Furthermore, we are not passive processes, tossed hither and yon through life by a blind watchmaker and the dictates of our selfish genes, but active participants, helping to shape our destinies as we go along. We are to some significant degree "self-created." Rose writes: "The central property of life is the capacity and necessity to build, maintain and preserve itself, a process known as autopoiesis" (p. 18). On page 6, he opines, "We are...the products of the constant dialectic between the biological and the social."

Rose also points out that our ability to perform experiments in the real world is limited; how even the most dedicated and thorough scientist in the field can only hope to observe a sampling of the behavior of the animals he or she is watching; how the variables in the real world are so very, very many; and how our attempts to control them can actually result in a falsification of the environment we want to observe. He argues convincingly that the hard sciences, especially physics, have yielded to reduction simply because they are not anywhere near as complex as biology.

Consequently I was very impressed with this book for the first 174 pages or so. Then came the chapters on evolution. Suddenly Rose unaccountably loses his objectivity and his reasoned tone and starts inventing straw men, one he calls "sociobiology" and puts these macho words in its mouth: "Males and their sperm compete, females and their ova quiescently await their fate" (p. 198).

Oops, have I picked up the wrong book? Could this be some rad fem polemic intent on winning some political point? This claim that sociobiologists think that females "await their fate" is particularly startling since on the previous page Rose writes that "Darwin's view was that, by and large, it is the female of the species that does the choosing." Rose then mocks the idea that there might be universal standards of beauty (that would be politically incorrect, no doubt). But the truth is, that while people can and do differ in details, a young, healthy, well-proportioned ("symmetrical," if you will) woman is recognized as attractive in any culture that I have ever heard of. On the next page (199) he makes fun of the idea that human females may choose males with resources ("the Porsche and the Rolex") adding that "wealth is no measure of genetic fitness...nor is there much evidence that its possession results in a greater number of offspring."

Rose knows this is fatuous. It is universally recognized that females across cultures prefer men of means. Why would a reasonable woman, given a choice, choose a poor, ineffective, unsuccessful man, to one who has the ability to help her provide for her children? Rose allows that sexual selection "may be— probably is—an important mechanism...but...we should not let its enthusiasts blind us to the more obvious explanations for the complexity of human sexual arrangements." Those "enthusiasts" are, one presumes, misguided sociobiologists. (Perhaps Rose would like to be regarded as a biology "enthusiast.") And just what are those "more obvious explanations"? Rose does not say.

He goes on to altruism but doesn't mention the handicap principle from *The Handicap Principle: A Missing Piece of Darwin's Puzzle* (1997) by Amotz and Avishag Zahavi, which I recommend that he read. This principle accounts for some acts of altruism by showing that such acts are the advertising of one's ability to others, in particular members of the opposite sex, and are therefore adaptive. He closes the section with a story about "two human sociobiologists" who thought that they had demonstrated that parents who both voted Conservative were more likely to send their child to a private school. Oh boy, and I might find two biologists who voted Liberal who were therefore more likely to send their child to a non-denominational school. As Rose's esteemed colleague, Steven Jay Gould likes to say, "So what?"

Rose begins the next chapter by asserting that "ultra-Darwinism" has "a metaphysical foundation" that includes the premise that "the purpose...of life is reproduction." I don't know who these "ultra-Darwinians" are but most experts on evolution tend toward the idea that "purpose" is an anthropological idea inconsistent with evolutionary theory.

So why is this book reasonable and fair three quarters of the way through and then suddenly we come upon prejudicial attacks against nonexistent bogeymen? It's the same old problem: a personal agenda. No matter how expert one may be, if the subject strays to the area of one's prejudices there is the chance that one may suddenly become as fair and objective as a radio talk show host.

What Rose wants to save us from is determinism, particularly genetic determinism. He thinks that determinism in biology or psychology may lead to the justification of some discredited ideas from eugenics. I don't agree. I think we can safely bury eugenics and such delusions as I.Q. and racial significance. The ghosts of the past are scary and we should be on watch, but we don't have to discredit the insights and accomplishments of sociobiology and/or evolutionary psychology by falsely associating them with those old, tired delusions.

Russell, Robert Jay *The Lemurs' Legacy: The Evolution of Power, Sex, and Love* (1993) ****
From the mother-daughter bond to the war system

This was published the same year as Matt Ridley's *The Red Queen*, and like *The Red Queen* is a classic in the relatively new science of evolutionary psychology. It is a little dated; for example Russell wasn't aware that altruism is advertising. We enhance our status and dominance in the society by appearing altruistic. And Russell's view of feminine sexuality is a little Pollyannaish. He does not mention the now well-documented female strategy of cuckolding a mate for a reproductive liaison with what she perceives is an alpha male. See, e.g., Baker, Robin. *Sperm Wars: The Science of Sex* (1996) or Diamond, Jared. *Why is Sex Fun? The Evolution of Human Sexuality* (1997) for the sobering revelations. What sets this book apart from others is Russell's thorough discussion of the war system as practiced by primates, while his work with lemurs allows him to go back further in time for his speculations. His style, like Ridley's, is lively and very readable.

Russell's premise is that we descended from lemurs (and from shrews before that) and that our psychology today can be better understood through an examination of lemur and other primate behavior. This is part of the basis for evolutionary psychology, the idea that we can better understand ourselves by studying the behavior of animals that are genetically close to us, especially animals similar to ones in our ancestry. Russell makes a strong case for this point of view while gently dismissing psychoanalytic theories. He writes: "Freud made the mistake of ethnocentrism by concluding that the behavior of *Homo sapiens* could be understood from studies of behaviorally-troubled patients within his own society." (p. 24)

On page 152 is perhaps Russell's main point, that "War evolved to displace in-group male aggression." On page 193 he adds, "War, for twenty million years, has served the needs of the ruling oligarchy above all other considerations." Those needs include killing off young males who represent not only a threat to the power of the oligarchy, but sexual competition. In fact, war can be seen as a pact between the ruling classes of one tribe and another: you kill off our excessive males and we'll kill off yours, and we'll both benefit.

I have to disagree with Russell, however, on riots, which he equates with war.

The riots in the cities are not like war; they are what will result if an enemy outside society cannot be found. Then the ruling classes themselves will become the enemy. One method of dealing with the violent dissatisfaction expressed in riots is ruthless suppression, as in totalitarian governments. Another is to ship the omega males off to war as in both totalitarian and democratic societies. A third method, employed in the United States today, is to put them into prison. We are simultaneously raising the price of the drugs that the dissatisfied are addicted to while imprisoning them when they attempt to buy these drugs or when they commit crimes to get money to pay for the drugs. It's a system that appears to be working. Perhaps it is better than the war system.

Russell sees the use of language as a way to lie, mislead and deceive. "Romance requires deception, most often self-deception." (p. 183) He adds: "...it has been estimated that the living English language contains no fewer than 300 euphemisms for the word "penis," a clear indication of our preoccupation with sex and our attempts to keep communications about that important subject private, imprecise, and obscure." (p. 187)

The book ends with a clarion call to save the earth's tropical forests, etc. presented with a heavy dose of pessimism. Russell's concern is that there are already far too many humans on the planet. On page 239 he complains about "Well-intentioned humanitarian groups [that] feed, clothe, and house surplus children." He adds (still p. 239) "why feed prolific human breeders when we know that soon we will not have enough food to feed all their children? ... Saved children become breeding adults who repeat their parents' mistakes."

I tend to agree with this, but I might ask him about those Malagasy dogs that the blurb on the jacket says he's so fond of. Does he feed them meat from cows bred on land that previously contained a tropical forest or from the flesh of whales harpooned in the North Pacific?

Russell's is a voice in the wilderness, and from his strident tone, he knows it. I am glad that somebody agrees with me that there are too many people on this planet. I just hope we can curb our appetite for reproduction before it is too late.

Singer, Peter *A Darwinian Left: Politics, Evolution and Cooperation* (2000) ****
Well worth reading

It was thought not too many years ago that the architects (so to speak) of the modern world were Marx, Darwin, Einstein and Freud. Now that the post-modern era is upon us, a reevaluation has been made and Marxist ideas have been largely discredited. Einstein has suffered a correction (from quantum mechanics), Freud has been reclassified as literature, and it is only Darwin's reputation that has survived unsullied.

Furthermore during this period the right has taken Darwin as its own, believing that the competitive biological nature of human beings as revealed by evolutionary biology is what leads to the inequalities that exist in human societies while justifying the war of one against all, etc.

But what Peter Singer is crowing about (and is the occasion for this lengthy essay/short book) is that the "red in tooth and claw" (Tennyson) interpretation of biological evolution that prevailed throughout the modern era is now coming under fire. No longer can biological evolution be seen as simply the strong taking advantage of the weak (a notion understandably obnoxious to the left). The larger truth now emerging from biology is that cooperation plays an important role in being fit and has, especially for humans, great adaptive value. It is becoming clear that Richard Dawkins's idea of the "selfish gene" is only part of the understanding, and that natural selection operates on groups through the individual, leading to an understanding that one (more cooperative) tribe may be selected over another, and that it is through cooperation within the tribe that Darwinian fittest may be most strongly expressed.

Now this is an idea that the left can appreciate. Consequently Singer's enthusiasm. Marx is dead, long live Darwin!

My problem with this intellectual enterprise is one that Singer points to on page 38, namely that we cannot form an argument from what *is* to what *should be*. Singer opines that we can instead through an appreciation of evolution gain "a better understanding of what it may take to achieve the goals we seek."

Beginning on page 31 with his second chapter, Singer compares behaviors across societies. This allows him to note which practices are universal or nearly so and which are highly diverse. The conclusion is that the more universal the behavior, the more it is a product of our biological nature and not a construct of society. To the extent that this process is valid, the information gotten is valuable. This is indeed one of the tools of evolutionary psychology that some people on the Darwinian left would like to discredit. They fear that an emphasis on our genetic endowment will work against our ability to nurture positive values and behaviors. They want nurture trumping nature.

However, in my opinion, the entire argument is passé and invalid. It is now generally understood in biology that nature gives us a predisposition to certain behaviors that develop in concert with our environmental experience so that our behaviors are an intimate product of both our nature and our nurture and cannot in any way be separated. The old "nature vs. nurture" debate is now seen as based on a false dilemma.

Also, it should be appreciated that today's scientific understanding of human nature as derived from biology, genetics and kindred disciplines, is just that, today's understanding, and as such is tentative. Consequently any oughts, shoulds, etc. drawn from such an understanding—even if such a practice were logically valid—would also be of a provisional nature.

Having said all this, I want to note that Singer's argument is well presented and his prescription for a Darwinian left in Chapter 5 well worth reading. If adopted it would work toward relieving the left from its fear of what evolutionary psychology is discovering about human beings. As Steven Pinker (not exactly a leftist) cheerfully notes, "Singer challenges the conventional wisdom that a recognition of human nature is incompatible with progressive ideals..."

He does, and indeed Singer demonstrates that the discoveries of evolutionary biology can be completely compatible with the traditional values of the left. This is an important understanding, since evolutionary biology is not going to go away, nor are its discoveries. We must learn to live with who and what we are without necessarily condoning our less attractive tendencies or attempting to sweep them under the rug.

Bottom line: the opening chapter which concentrates too much on the well-known Marxist delusions and the Soviet doublethink might well be skipped.

The meat of Singer's essay begins with Chapter 2, and works very well by itself.

Smith, David Livingstone *The Most Dangerous Animal: Human Nature and the Origins of War* (2007) *****
War from an evolutionary psychological point of view

Once upon a time we were little australopithecine animals living in mortal fear of the great carnivores as we tried to steal bones from their kills, sleeping at night in trees where great snakes and huge eagles treated us as prey. Then some time later we grew larger and smarter and begin to ward off the carnivores with sticks and stones and group cohesion. And then there came the day when we became the most feared predator of them all.

This little history, according to the lengthy and perceptive analysis in this most engaging book, sheds important light on why we wage wars and kill with such ferocity.

"The Most Dangerous Animal" is us. We have guns and walls and locks to protect us not from lions and tigers but from each other. But to gain the right ferocity and the sheer bloodlust needed to defeat our human enemies, we had to turn them into beast and vermin and other nonhuman creatures because, simultaneously with our ability to kill, we had a mental module that urged us not to kill our kind. Therein lies, according to Professor Smith, who is both a philosopher and a psychologist, the terrible dialectic that is the human mind as warrior. For the tribe to survive it had to be able to stir its young men to a killing rage like chimpanzees tearing a strange chimp to bits with their teeth and bare hands. But at the same time, this violent ferocity must not be turned upon family, friends and other members of the tribe. And so these two assortments of mental neurons (mental modules) exist simultaneously in the human brain, and depending on circumstances lead us to brotherhood or to genocide.

The question that confronts us today is will we always have war? When I was an undergraduate I argued against the affirmative with others and in particular with one of my psychology professors. In the final argument it came down to the definition of war. If war is any violence of humans against humans, then, yes, war will never end until our nature changes, possibly through some kind of biological engineering. But if war is tribe against tribe, nation against nation, then it is possible that through the rule of law imposed internationally upon all people, war may end. Possibly. Smith is pessimistic, and I can say—no longer an undergraduate—that unless human nature changes, there will always be disputes that sadly cannot be settled in any other way. War is "politics by other means."

Smith defines war as "premeditated, sanctioned violence carried out by one community (group, tribe, nation, etc.) against members of another." (p. 16) He recalls the work of Jane Goodall and others who observed chimpanzees carrying out "raids" against other chimps in a purposeful way that is very much like humans going to war. Since we are genetically very much like chimpanzees, their behavior suggests a common inherited source of warlike violence. But Smith also points to the bonobos, the smaller chimps who practice what can only be called "love not war"—or at least "sex not war." They too are our close cousins. And how like caricatures of the human left-right political dichotomy are the two types of chimp! I think what we need to understand is that those who believe in the war system and those who do not, come by their beliefs genetically. Their beliefs are ingrained. And in many of us both beliefs are held simultaneously leading to cognitive dissonance.

What we do, as Smith so painstakingly demonstrates, is we lie to ourselves. We practice self-deception to an amazing degree. Smith even argues that self-deception is adaptive in the Darwinian sense. He cites biologist Robert L. Trivers as arguing that self-deception is adaptive because it is easier to fool others when we have first fooled ourselves. (p. 126) Furthermore, how do we avoid guilt and self-loathing after killing another human being in cold blood on the battlefield? Or better yet, how do we get our young men to do this killing? We convince ourselves first, and then them, that our adversaries are monstrous vermin, that they are subhuman, that, although they have a human form, they lack the "essence" of being human. Smith gives many examples of people from ancient times to the present day as doing exactly this. The prelude to genocide is the dehumanization of others.

But this book is about more than the war system. Professor Smith demonstrates a profound understanding of human psychology in other areas as well. His take on consciousness is one of the best I have ever read. He writes: "...it is a mistake to imagine that there is something in the brain corresponding to our notion of consciousness. Consciousness is not a thing inside the brain rubbing shoulders with the anterior cingulated gyrus or tucked away discretely behind the amygdala. Consciousness—if one wants to use this slippery term at all—is something that the brain does. The fact that the word "consciousness" is a noun half-seduces us into thinking of it as a thing. The word 'consciousness' should have a verbal equivalent: we should be able to say that the brain is 'consciousnessing'." (p. 104)

Actually we do have such a verbal equivalent. It is "perceiving." Consciousness is perception, but perception writ large, including partial perception of our inner states and our mental activities, and the feelings that come from our emotions, as well as what has happened, is happening, and is likely to happen, around us. This is in addition to the perception that comes from the "third eye"—the mind. This perception, at which we are the planet's clear leaders, combines knowledge from perceptions about things past and present, about things seen and heard and told about, and puts all that information together in a grand mental perception about what has happened, is happening or is to come.

Stanford, Craig *Significant Others: The Ape-Human Continuum and the Quest for Human Nature* (2001) ****
Flawed, but definitely worthwhile

University of Southern California anthropologist (and primatologist) Craig Stanford's thesis in this attractive but somewhat breezy (and politically considered) book is that the difference between humans and apes is one of degree and not of kind. That is why the word "continuum" is used in the title.

I agree with his thesis, and I think he does a great job of making the case. His prose is readable and his enthusiasm is genuine. However there are some problems. In attempting to walk the tightrope of political correctness while conveying to the reader what he has learned as a scientist, Stanford sometimes slips into a fuzzy and inexplicit expression.

To begin with (p. 16) he contends that if women "crave" men with resources (he is attempting to answer David Buss, et al.) it is "mainly in patriarchal societies in which they must depend on men to obtain resources and power for them." This is gratuitous because, as Stanford himself notes on page 147, "Human societies are, political correctness notwithstanding, universally patriarchal." Whether women would behave differently if the societies were matriarchal (or otherwise) is unknown. Citing an isolated society in special circumstances that is matriarchal really does not prove the general case, although it does point to a range of possibilities, and that is good. However it is ingenuous to pretend that women are not looking for resources in a mate if they can find them. Why would a reasonable woman, given a choice, choose a poor, ineffective, unsuccessful man, to one who has the ability to help her provide for her children?

In the same paragraph, Stanford contends that the "old adage about 'what women want' should more accurately be phrased as 'what women can realistically hope to achieve in their cultural context'." In the first place, it's not an adage, it's a joke or a lament, and it's a question, "What *does* a woman want?" The original is lost in the prehistory, but Ernest Jones attributed these words to Freud: "The great question...which I have not been able to answer, despite my thirty years of research into the feminine soul, is 'What does a woman want?'" This is probably the source that Stanford had in mind—and, by the way, this is a question that evolutionary psychology has largely answered, much to the dismay of those who would prefer to keep the mystery. In the second place it is *not* enough to merely say "in their cultural context." There is a biological context as well, exemplified by nine months of being pregnant, and several years of intense maternal responsibility that is fundamental to all cultures that can't be explained away as something from the patriarchy unless you believe that human biology itself is patriarchal.

There is also Stanford's summary dismissal of evolutionary psychology in Chapter 8 to consider, a strange dismissal since part of his title is "the Quest for Human Nature" (from the study of primates), which is one of the ways that evolutionary psychologists work. (He is actually being an evolutionary psychologist himself but apparently doesn't know it.) Evolutionary psychology should be compared with other psychologies, say, psychoanalytical theory, or behaviorism, and not to, e.g., biology.

It's important to add that the work of anthropologists is no more scientific or rigorous (or less so) than the work of evolutionary psychologists, as can be demonstrated from reading this book. For example on page 129 Stanford tells a story of seeing the low-ranking chimp Beethoven make a sexual display through a cluster of chimpanzees. He writes: "This enraged the alpha, Wilkie, who chased Beethoven off into the thickets, whereupon Beethoven circled around and came back to mate with an eager female before Wilkie realized what was happening."

Stanford uses this as an example of planned deceptive behavior in chimps, but whether Beethoven displayed foresight or just got lucky is unclear. To be picky I could also point out that "enraged" and "realized" are anthropomorphic projections of Stanford's lively mind and not something that could be tested scientifically (which is the essence of his criticism of evolutionary psychology on page 134).

Yet, Professor Stanford understands that social scientists today are mightily constrained by a postmodern culture in academia that demands politically correct findings first, and scientifically persuasive findings only if they are in agreement with the PC party line. He writes, "Some of this sentiment [not admitting "essential cultural commonalities"] reflects anthropologists' political burden of favoring the cultural underdog at all costs. Postmodernism's purpose has become a vehicle, in part, to give meaning to identity politics in the battle of the oppressed against the perceived enemy, the white male elite." (p. 146)

There is a lot more worth discussing in this book. (I wish I had more space.) The chapter on what it's like in the field (Chapter 12) is vivid and compelling; and in the concluding chapters we can see that Stanford is a scientist who cares passionately about the great apes and their environments. He is also a man who can communicate what he knows to a general readership as long as he avoids the trap of imagining that there is a political censor sitting on his shoulder as he writes. The truth will out, and the educated public that reads books written by professional scientists is much more sophisticated than is sometimes supposed.

Tattersall, Ian *The Monkey in the Mirror: Essays on the Science of What Makes Us Human* (2002) *******
Lively, interesting, but not entirely objective

There are eight essays. The first one, "What's So Special about Science?" explains to the general reader what science is and what it isn't. One of the points Tattersall makes is that "scientists are emphatically *not*...steadily building up a picture of *the truth*." (p. 8) Instead, "the solution of one scientific problem regularly leads to the identification of others...," so that the "successful climbing of an intellectual summit has always revealed new peaks beckoning beyond." (p. 7) Tattersall adds on page 30 that "science is a system of provisional knowledge" that "does not seek to understand ultimate causation..."

In the second essay, "Evolution: Why So Misunderstood?" Tattersall argues that many people think that science is an "authoritarian" system that "produces axioms that are unchanging for the ages." Consequently science runs afoul of other authoritarian (read: religious) systems that feel threatened from without. I think this is a good argument, but I think it is also the case that evolution is so incredibly complex that it is not easy to understand or appreciate. Tattersall writes on page 29 that "The notion of evolution is, after all, a pretty simple one..." Yes, the notion is relatively simple, so simple that Thomas Henry Huxley exclaimed, "How very stupid of me not to have thought of that!"; but after the notion comes an amazing, really stupefying mass of complexities. In truth very few people really understand even the basics of evolution beyond the initial idea. And within the ranks of the experts there are endless arguments.

The next two essays, "The Monkey in the Mirror" and "Human Evolution and the Art of Climbing Trees" reminds us that monkeys cannot recognize their reflections in the mirror, but that we and the chimpanzees can. Here Tattersall gives us his view on consciousness and its evolution based primarily on evidence from the fossil record. Tattersall's position is highly tentative and emphasizes how little we really know.

In the next chapter, "The Enigmatic Neanderthals" Tattersall sums up what we know about the Neanderthals and what happened to them. I was interested to notice that his ideas are not far removed from those presented fictionally many years ago by William Golding in his novel *The Inheritors*, namely that we somehow, probably by violent force, brought the big and strong Neanderthal to extinction. Golding (and the evolutionists of his day) emphasized the murderously deceptive mentally agile of homo sapiens as the decisive factor while

Tattersall believes the jury is still out on exactly why they disappeared. Another possibility is that our diseases killed them, but most probably it was a combination of factors that led to their demise.

The final three essays attempt to account for our "humanity" and where we might go from here. Tattersall makes the very important point that speciation can only occur in isolated populations; consequently our population being both six billion strong and in full interaction, there is little prospect, barring catastrophe, for our further evolution. He writes, "the trend is exactly the opposite to what is required for any meaningful evolutionary change..." (p. 190) Of course there *is* cultural evolution to consider, a subject that Tattersall understandably does not address in a collection of essays on biological evolution. For some ideas about what may become of us through cultural evolution see, Pierre Baldi's *The Shattered Self: The End of Natural Evolution* (2001) or Ray Kurzweil's *The Age of Spiritual Machines: When Computers Exceed Human Intelligence* (1999).

I have a couple of bones to pick (if you will) with Tattersall. First there is the little matter of attributing to Shakespeare the famous phrase "nasty, brutish and short" (p. 170) in describing human life in the wild. As most political and social science majors know, the phrase is from the *Leviathan* (1651) by Thomas Hobbes. Also, even appreciating that Tattersall is taking a causal tone here and forsaking the sort of scientific rigor and fairness shown in more academic tomes, there is no excuse for this from page 50, "Much of the discussion of *adaptations* doesn't even have...[some minimal basis in empirical fact], as we will discover when we look at the arrogant pseudo-science of evolutionary psychology in a later essay."

Tattersall does take a quick look at evolutionary psychology in a later essay, but in no way does he support his dismissive charge or even indicate just who or what it is that is "arrogant." It is especially distressing to note that Tattersall throughout this book again and again prefaces his suppositions with words like "It seems reasonable to conclude..." or something similar (see for example, pages 96 and 98) yet he denigrates evolutionary psychology for no greater crime than drawing reasonable conclusions. He writes that his argument with evolutionary psychology is in its undue reliance of genetics (beginning on page 170), but actually the power of evolutionary psychology comes not from assigning behaviors to genes, but from drawing insights into our behavior from the process of evolution and from the behavior of other animals. From that evidence, it is reasonable to conclude any number of things, and they are worth noting, even if there is no way they can be proven, any more than a host of Tattersall's conclusions about human evolution can be proven.

Furthermore he accuses (again unnamed) evolutionary psychologists of "defending rape as an adaptive behavior..." (p. 178). I personally know of no evolutionary psychologist who would do such a thing. Why doesn't Tattersall name one? What evolutionary psychology is saying is not that rape can be defended. It can't. But that there are evolutionary reasons for its existence. This is quite a different statement. There are evolutionary reasons for murder, etc., but in recognizing them, that does not mean we are "defending" them or are in any way in agreement with them anymore than Tattersall is in agreement with what we presumably did to the Neanderthal.

Wade, Nicholas *The Faith Instinct: How Religion Evolved and Why It Endures* (2009) *****
Chronicles a paradigm shift in the way we view religion

Nicholas Wade, who also wrote the very fine *Before the Dawn: Recovering the Lost History of Our Ancestors* (2007) (see my review at Amazon), argues most convincingly here that religion, our sense of spirituality, and our moral instinct have been hardwired into our brains by the evolutionary process. This book, supported in part by the Templeton Foundation, is the first of its kind to put together the body of evidence that accounts for the fact that religion has been part of every known human society while explaining why.

Is religion adaptive in an evolutionary sense? is the first and most important question to be answered. The fact that religion is universal strongly suggests that it is. But until recently this idea was rejected by most biologists including some heavy hitters such as George Williams, Richard Dawkins and Steven Pinker. But, as Wade points out, Dawkins and Pinker in particular may have missed the boat because of personal biases. Wades writes that their opposition "seems to be driven less by any particular evidence than by the implicit premise that religion is bad, and therefore must be non-adaptive." (p. 67)

Moreover, Williams and Dawkins have been against the idea that religion is adaptive because of their belief that natural selection operates primarily at the level of the individual. For religion to be adaptive in the Darwinian sense, it helps a lot for selection to operate at the level of the group. Wade shows that biologists such as David Sloan Wilson and Edward O. Wilson, not to mention Darwin himself, support the idea of group selection. Wade presents Darwin's argument from the *Autobiography* (see page 68) that tribes who had members who were ready to sacrifice themselves for the good of the tribe would help their tribe prevail over other tribes without such people. Williams and others came to differ with Darwin by arguing that free-loaders and cheaters only interested in promoting their own genes would out-reproduce the do-gooders. This opinion has held sway in evolutionary biology for a long time, but that is changing. Wade quotes David Sloan Wilson and Edward O. Wilson as putting it this way: "Selfishness beats altruism within groups. Altruistic groups beat selfish groups. Everything else is commentary." (p. 70)

But how is religion adaptive? Why should those tribes that were religious have out-competed those that were not? Where are those non-religious tribes? The answer is there aren't any. The assumption is that they were driven to extinction by the religious tribes.

Just what is it about religion that confers upon its practitioners such a huge evolutionary advantage? The answer in a word is warfare. The intimate relationship between human warfare and religion is really the crux of the matter. As warfare became more important among human groups competing for scarce resources a greater premium was placed on winning. What religion does so very well is make the tribe more cohesive than it would otherwise be.

One of the most interesting things about religion as revealed in this book is that religion came before language! How can that be? Wade explains that in the most primitive societies, the basis of religion is communal, rhythmic singing and dancing. This singing and dancing can be seen to draw the members of the tribe closer together so that they can act as one with less fear of danger as they are strengthened by the cohesiveness of the group. People could dance and follow rhythms and perhaps sing before they could use syntactic language. We see many animals, especially birds, that perform elaborate dances. Hominids, being social creatures would dance en mass not so much to be sexually selected (although that too no doubt) but to strengthen their ties within the group.

But this ecstatic expression of religion cuts both ways. In historic times religion has become hierarchical, the rituals have become more sedate, and the basis of group membership is based not on ecstatic communal expression but more on shared beliefs. In fact some religions have banned dancing. Wade suggests that this is because the power of the leaders of these modern religions can have their authority threatened by deeper and more immediate appeals to emotion. This might be what is happening in Latin America today with membership in the Catholic Church shrinking while membership in the more demonstrative Protestant churches with singing and even speaking in tongues gaining adherents.

In the latter part of the book Wade traces the birth and growth of various religions including especially the three monotheistic religions from the Middle East. He doesn't see religion as the cause of wars per se, only as a very nice tool for being successful in wars! Finally he looks at the future of religion. He hints at a need for religions that are more in tune with the modern world. Beyond that he does not go.

All in all an excellent book that deserves a wide readership.

Wilson, David Sloan *Evolution for Everyone: How Darwin's Theory Can Change the Way We Think about Our Lives* (2007) *****
Evolutionary theory as a guide to life

"The most extraordinary fact about public awareness of evolution is not that 50 percent don't believe the theory but that nearly 100 percent haven't connected it to anything of importance in their lives." (p. 315)

This is a bit curious, but when you consider that Edward O. Wilson's *Sociobiology* was published only 34 years ago, and further that evolutionary psychology has only recently made its way into the curriculum of our university psychology departments, it is understandable. For my part, like David Sloan Wilson (son of Sloan Wilson who wrote a couple of fiction bestsellers in the 1950s, *The Man in the Gray Flannel Suit* and *A Summer Place*), I took to the application of evolutionary ideas to my life the way a duck takes to water. But the overall public awareness and acceptance has lagged, in part due, as Wilson explains, to the failure of the larger academic community to incorporate evolutionary ideas and findings into their fields of study.

That is changing fast with evolutionary medicine, evolutionary psychology and other scientific approaches now established fields of study. What David Wilson hopes follows is an awareness of evolutionary ideas and principles in the social sciences and the humanities, which is one of the reasons he wrote this book which grew out a class he taught to undergraduates.

The essence of evolutionary thought as applied to our daily lives is to ask the question, how does such and such a behavior or such and such an idea relate to the way evolution works? For example, not so long ago we were urged to drink lots of water every day (probably from studies funded by bottled water companies!). But if you think about the human experience in the Pleistocene in what is called the Environment of Evolutionary Adaptation (EEA) you might ask yourself, how was it possible for humans to drink so much water? Clearly humans would develop an ability to function very well, even optimally, without having to drink so much water, which in those days and climes would have been difficult to do safely. Consequently, doing this "thought experiment" I began to doubt the necessity to drink so much water. And lo and behold it came down from newer studies that actually humans don't really need to drink so much water! David Sloan Wilson gives a number of other examples of evolutionary thinking that has helped us to better understand ourselves and our place in the world and our communities. He is very strong on the idea of cooperation as an adaptive force in evolution, especially human evolution.

One of the ideas that most impressed me is his recognition of the arms struggle between society and the "selfish" individual. Some old-line evolutionists are loath to accept altruism and other seemingly selfless behaviors that benefit the tribe or larger groups as adaptive (other than through kinship) since the genes that code for such behavior would be easily overrun by genes from individuals looking out only for themselves. But what I think is overlooked is the human ability to spot these cheaters and keep them in check or to kick them out of the tribe or worse. Wilson makes the very interesting point that gossip is part of this process. Through gossip a society "maintains a dossier of information on every member and quickly detects social failings." (p. 160). Sociopaths don't fare well in communities in which everybody knows everybody else. But of course gossip doesn't work well, and a sociopath can flourish, where almost everyone is a stranger to one another, which is usually the case in our big cities. This lack of communal checks explains in part why there is so much crime in our cities.

Another interesting and fundamental idea is what Wilson calls "dancing with ghosts." The idea is that the adaptations we made during the EEA in some cases no longer apply effectively to the current environment. Thus the very nice ability to efficiently put on fat when large amounts of sugar, carbs and fats are temporarily available worked well in the prehistory when the dearth of winter or the dry season was to come; but in today's world of supermarkets and a MacDonald's on every corner, this ability has become a detriment leading to obesity and chronic disease. Many people in the West are dancing with the ghosts of "eat your fill when it's available." This predictive adaptive response (PAR) is no longer adaptive. Wilson gives some other examples relating to pronghorn antelopes that still "flee with amazing speed and endurance from predators that no longer populate the American plains" and baby sea turtles that mistake the lights of the city for the moon shining off the ocean and crawl in the wrong direction. (pp. 52-53)

Wilson also argues convincingly for the idea that life in the ghetto is more dangerous than say life in the suburbs because young people in the ghetto must take greater chances in order to be gain status and wealth. For a person like David Sloan Wilson to risk his life for some status gain would be foolish since he is going to gain enough wealth and status to be successful because of his many social and economic advantages. For a guy in the ghetto, it is sometimes worth the risk (or so it seems) to fight another at the drop of an insult because of the gain in status that can lead to better mating opportunities and a greater command of turf. When the environment is "unstable" and "life expectancy" is "low," a good strategy is to "take care of immediate needs and reproduce early." When you have a "stable environment and high life expectancy" on the other hand, you should "plan for the long term, including delayed reproduction."

There is also a lot in this book about religion from an evolutionary point of view, which I don't have space to go into, some of it based on Wilson's earlier book *Darwin's Cathedral* (2002).

Wilson, Edward O. *On Human Nature* (1978) *****
Without euphemism

On reading this again after a couple of decades, I am struck with how brilliantly it is written. The subtlety and incisiveness of Wilson's prose is startling at times, and the sheer depth of his insight into human nature something close to breath-taking. I am also surprised at how well this holds up after all these years. There is very little in Wilson's many acute observations that would need changing. Also, it is interesting to see, in retrospect, that it is this book and not his monumental, *Sociobiology: The New Synthesis* (1975), that continues to serve as an exemplar for later texts. For example, Paul Ehrlich's recent book on evolution was entitled *On Human Natures* (2000), the plural in the title suggesting that it was written at least in part as a reaction to Wilson. I also note that some other works including Matt Ridley's *The Red Queen: Sex and the Evolution of Human Nature*.(1993), Robert Wright's *The Moral Animal: Evolutionary Psychology and Everyday Life* (1994), and more recently, Bobbi S. Low's *Why Sex Matters: A Darwinian Look at Human Behavior* (2000), are organized intellectually in such a manner as to directly update chapters in Wilson's book.

On Human Nature was written as a continuation of *Sociobiology*, greatly expanding the final chapter, "Man: From Sociobiology to Sociology." In doing so, Wilson has met with reaction from some quarters similar to the reaction the Victorians gave Darwin. Wilson's sociobiology was seen as a new rationale for the evils of eugenics and he was ostracized in the social science and humanities departments of colleges and universities throughout the United States and elsewhere. Rereading this book, I can see why. Wilson's primary "sin" is the unmitigated directness of his expression and his refusal to use the shield and obfuscation of politically correct language. Thus he writes on page 203, "In the pages of *The New York Review of Books, Commentary, The New Republic, Daedalus, National Review, Saturday Review*, and other literary journals[,] articles dominate that read as if most of basic science had halted during the nineteenth century." On page 207, he avers, "Luddites and anti-intellectuals do not master the differential equations of thermodynamics or the biochemical cures of illness. They stay in thatched huts and die young."

In the first instance, he has offended the intellectual establishment by pointing out their lack of education, and in the second his incisive expression sounds a bit elitist. But Wilson is not an elitist, nor is he the evil eugenic bad boy that some would have us believe. He is in fact a humanist and one of the world's most renowned scientists, a man who knows more about biology and evolution than most of his critics put together.

I want to quote a little from the book to demonstrate the incisive style and the penetrating nature of Wilson's ideas, and in so doing, perhaps hint at just what it is that his critics find objectionable. In the chapter on altruism, he writes, "The genius of human sociality is in fact the ease with which alliances are formed, broken, and reconstituted, always with strong emotional appeals to rules believed to be absolute" (p. 163). Or similarly on the next page, "It is exquisitely human to make spiritual commitments that are absolute to the very moment they are broken." Or, "The genes hold culture on a leash. The leash is very long, but inevitably values will be constrained in accordance with their effects on the human gene pool" (p. 167). He ends the chapter with the stark, Dawkinsian conclusion that "Morality has no other demonstrable ultimate function" than to keep intact the genetic material.

In the chapter on aggression, he posits, "The evolution of warfare was an autocatalytic reaction that could not be halted by any people, because to attempt to reverse the process unilaterally was to fall victim" (p. 116). On the next page, he quotes Abba Eban on the occasion of the 1967 Arab-Israeli war, "men use reason as a last resort."

In the chapter on religion, he argues that the ability of the individual to conform to the group dynamics of religion is in itself adaptive. As he avers on page 184, "When the gods are served, the Darwinian fitness of the members of the tribe is the ultimate if unrecognized beneficiary."

It is easy to see why some people might be offended at such a frank and penetrating expression. But one of the amazing things about Wilson is that he can be bluntly objective about humanity without being cynical. I have always found his works to be surprisingly optimistic. He has the ability to see human beings as animals, but as animals with their eyes on the stars. In the final chapter entitled, "Hope," Wilson presents his belief that our world will be improved as scientific materialism becomes the dominate mythology. Note well this point: Wilson considers scientific materialism, like religion and the macabre dance of Marxist-Leninism, to be a mythology. His point is that there is no final or transcending truth that we humans may discover; there is no body of knowledge or suite of disciplines that will lead us to absolute knowledge. There are only better ways of ordering the environment and of understanding our predicament. He believes that toward that end scientific materialism will be a clear improvement over the religious and political mythologies that now dominate our cultures.

No one interested in evolutionary psychology can afford to miss this book, even though it was written the 1970s. It is a classic. Anyone interested in human nature (yes, one may profitably generalize about human nature, as long as one understands what a generalization is, and appreciates its limitations) should read this book, one of the most significant ever written on a subject of unparalleled importance.

Wrangham, Richard, and **Dale Peterson** *Demonic Males: Apes and the Origins of Human Violence* (1996) ****
Imagine the Victorians reading this book

This book is about bonobos and chimpanzees, gorillas and orangutans and their social behavior as it relates to ours, how we are similar, how different. Chimpanzees are surprisingly violent and make gang raids on neighboring tribes and kill (with their bare hands, by the way) if the raiding party has a big advantage. We and the chimps broke off from a common ancestor about five or six million years ago. We went out of the forest and onto the woodlands and the savannas. We learned to dig up and eat roots and of course bone marrow from kills. This is how we survived the loss of the forest and the recurring dry seasons.

We are a little less closely related to gorillas and orangutans, but the point the authors are making is we are the fifth ape, and it is valuable to study how the other apes behave so as to gain insight into ourselves. While the authors seem to lay the problem of human violence squarely at the feet of males, it is allowed—albeit only briefly and incidentally (p. 239-240)—that women choose these demonic males through sexual selection, and ultimately the problem of male violence is a human problem.

What is especially interesting here is the thorough examination not only of the violence practiced by apes, but of their differing sexual practices: gorillas form harems with a single silver back male getting most of the reproductive tries, while orangutans live alone and the males often engage in rape. In contrast the bonobos are so frequently and openly sexual that genital rubbing is a way of greeting while the father of the little ones could be any one of the males.

This is evolutionary psychology with a wary eye on political correctness. I note that Edward O. Wilson does not appear in the bibliography but Naomi Wolf and Andrea Dworkin (for example) do. In fact, this book is something of a blatant attempt to make evolutionary psychology palpable to women. The authors even have a category they call "evolutionary feminism" represented by "writers like Patricia Gowaty, Sarah Hrdy, Merdith Small, and Barbara Smuts" united in their opposition to "the patriarchy" (p. 124). This is all to the good of course because a thorough going understanding of human nature will lead us all to the inescapable conclusion that blaming one sex for the human problem of violence really misses the profound truth of sexual equality. The authors even suggest (p. 125) that "If all women followed Lysistrata's injunctions and refused their husbands, they could indeed effect change."

If. By the way: ugly dust cover.

Wright, Robert *The Moral Animal: Evolutionary Psychology and Everyday Life* (1994) *****
A classic worth a second look and an update

Although published in 1994, a long time ago in the rapidly developing science of evolutionary psychology, Robert Wright's seminal book remains an excellent introduction to the subject. The text crackles with an incisive wit that says, yes we're animals, but we can live with that. The discussion is thorough, ranging from a rather intense focus on Charles Darwin and his life through the sexist and morality debate occasioned by the publication of Edward O. Wilson's *Sociobiology* in 1975, to the rise of the use of primate comparisons fueled by Jane Goodall's instant classic, *The Chimpanzees of Gombe: Patterns of Behavior* (1986). Wright has some rather serious fun with human sexual behavior as seen from the perspective of evolutionary psychology, but he spends even more time worrying (to no good effect, in my opinion) about altruism and the shaky concept of kin selection. The title is partly ironic, since much of the material suggests that we are something less than "moral." The "Everyday Life" in the title is an allusion to Freud (*The Psychopathology of Everyday Life*, 1904) who makes a dual appearance in the text, first as a kind of not-yet-illuminated precursor to modern Darwinian thought, and second as the reigning champ of psychology that evolutionary psychology is out to dethrone. (See especially page 314.)

What's exciting about evolutionary psychology is that for the first time psychology has a firm scientific foundation upon which to build. But it's a tough subject for some people, I think, mainly because they confuse "is" with "ought." The discoveries of evolutionary psychology about the differing reproductive strategies of the sexes offend some people in the same way that Darwin's insight about our kinship with (other) animals offended the Victorians. Evolutionary psychology shows us that men lie, cheat and hustle relentlessly for sex, while women manipulate available males into caring for their offspring, and if possible for children fathered by other males. Insights like these are seen by some as immoral imperatives, when in fact they are amoral statements of factual observation. What "is" isn't necessarily the same thing as what ought to be. And really, we shouldn't blame the messenger.

Where Wright's book especially shows its age is in trying to explain altruism. He wasn't aware of the handicap principle developed by Amotz and Avishag Zahavi in their exciting book, *The Handicap Principle: A Missing Piece of Darwin's Puzzle* (1997) which nicely explains "altruism" (it's an advertisement of fitness) and a number of other evolutionary conundrums, including Wright's question on page 390, "Why do soldiers die for their country?" Additionally on pages 68-70, where Wright attempts to account for female cuckoldry, he gives three reasons, but seems uncertain of the most important one, which is that a woman, once established in a secure pair-bond will sometimes seek to upgrade the genetic input by having a clandestine fling with what she sees as an alpha male. Also Wright's attempt to account for homosexuality (pages 384-386) stumbles over itself in trying to be politically correct while missing the major point that homosexuality facilitates male bonding and therefore is certainly adaptive since male coalitions increase each member of the coalition's chance of securing females. It fact, Wright misses the whole concept of male bonding. There's not even an index entry for it.

These observations are not to be taken as criticisms of the book since Wright was writing before knowledge of some of these ideas became widespread. *The Moral Animal* remains an outstanding opus and one that has helped introduce a large readership to the power and efficacy of evolutionary psychology, a scientific approach to psychology that will, I believe, replace the old paradigms currently holding sway in our universities. Of course this will only happen when the old behaviorists, and cognitive and psychoanalytic stalwarts...retire.

I would like to see Wright revise this book in light of the many discoveries made during the nineties and reissue it. His readable and engaging style would make the update fun to read.

Zuk, Marlene *Sexual Selections: What We Can and Can't Learn about Sex from Animals* (2002) ****
Reconciling feminism and evolutionary biology

University of California, Riverside biology professor Marlene Zuk, whose specialty is insects, especially crickets, makes two main points in this modest volume. One, what is "natural" as observed in nature is not necessary right and should not be used as a guide for human society; and two, how we interpret the behavior of animals is colored by our biases, both anthropomorphic and male-gendered.

Professor Zuk writes from the avowed position of a feminist, although she makes it clear that she is not an "ecofeminist" nor does she agree with those feminists who believe that the exercise of science and "attempts to study the world are just culturally derived exercises relevant only in a certain social context." (p. 16)

In other words, Zuk wants to reconcile the ways of science, especially evolutionary biology, to feminists while pointing out to biologists that many of their preconceptions contain a male bias. She recalls a poem from A.E. Housman that includes the phrase "witless nature" which she takes as a cornerstone for her position. Nature "is not kind, not cruel, not red in tooth and claw, nor benign in its ministrations. It is utterly, absolutely impartial." (p. 15)

From this it follows (for most of us anyway) that we should not draw moral conclusions about how people should behave, nor should we form notions of what is "right" or "wrong" from observations of nature. This is a position that most professionals in evolutionary biology today appreciate, although this was not always the case, as Zuk is quick to remind us. She sees the antiquated notion of *scala naturae* (from Aristotle) which puts humans at the pinnacle of evolution as part of the reason for the errors of the past. Humans were seen as the positive norm, and to the extent that the behavior of other animals deviated from that they were inferior. Zuk also points to a "male model in biology" assumed by biologists (consciously or unconsciously), as an addition source of bias. She points to the idea that males are more aggressive than females as an example of an unwarranted preconception.

My experience (for what it's worth—I coached girl's basketball some years ago, and believe me the girls were *very* aggressive), and from what I know of aggressiveness theoretically, suggests that females are indeed just as aggressive as males in going after what they want. The reason that women use violence (a kind of aggressiveness) less than men do has to do with social conditioning of course, but also with the fact that a woman's reproductive capability is seldom if ever enhanced by the use of physical force while a male may use force to his reproductive advantage. In the case of non-human animals I am thinking especially of male lions killing the cubs of another male to bring the female into estrus. In the case of humans I am thinking of human males using the spoils of war to gain access to females and to nurture their offspring. (I am *not* thinking of rape since that sort of unsocial, high-risk behavior seldom leads to successful reproduction; more often it leads to ostracization and an early demise for the rapist, a state of affairs that is not adaptive, not to mention infanticide and abortion.)

Zuk writes in a witty style that is easy to read. Her target readership is the non-specialist; indeed one gets the sense that she is addressing her undergraduate students. Politically speaking, she steers a middle course between the extremes of the sociobiological right and the socialist left, a fact underscored by the appearance on the cover of endorsements from Matt Ridley on the right, Patricia Adair Gowaty from the left, and Sarah Blaffer Hrdy from somewhere in the middle.

I would give a more ringing endorsement of this book were it not for the fact that there is virtually nothing new in Zuk's very agreeable presentation, and my lingering sense that a person who identifies herself as "feminist" biologist (instead of merely a biologist) is not entirely objective any more than the old guys from the patriarchy were. However, to be fair, at no place in the book does Zuk espouse anything close to a preference for the politically correct at the expense of scientific inquiry, as feminists sometimes do when the conclusions are not what they want. Zuk knows that to make science subordinate to what is politically and socially agreeable is to sacrifice science completely. Indeed, I see this as the profound central message of her book, and a reason to hope this book receives a wide readership.

Chapter Five

Evolution in General

Some of the books in this chapter are about evolution in the broadest sense; some others are on specific kinds of plants or animals; others are on ideas in evolutionary biology; and still others are on historical aspects of Darwinism, including disputes and even internecine politics; and finally there are some books about the personal reflections of some great biologists. Several of the books are by non-specialist or experts in allied fields, such as Howard Bloom and Keith E. Stanovich who give us insights into evolution from perspectives outside the discipline itself.

Barlow, Connie *The Ghosts of Evolution: Nonsensical Fruit, Missing Partners, and Other Ecological Anachronisms* (2000) *****
Who mourns for the mastodons?

The tusks that clashed in mighty brawls
Of mastodons, are billiard balls...
                    —from a poem by Arthur Guiterman

The exciting idea in this book is that there are trees that "lament" the passing of the mastodons and the other extinct mega fauna that once distributed their seeds. What animal now regularly eats the avocado whole, swallows the seed and excretes it far from the tree in a steamy, nourishing pile of dung? No such animal exists in the Western Hemisphere to which the avocado is native. (Barlow reports that elephants in Africa, where the avocado has been introduced, eat the avocado and do indeed excrete its pit whole.)

How about the mango with its pulp that adheres so tightly to the rather large pit? As Barlow surmises, such fruits were "designed" for mutualists that would take the fruit whole and let the pit pass through their digestive systems to

154

emerge intact for germination away from the mother tree. Note that the avocado pit is not only too large to pass comfortably through the digestive system of any current native animal of the Americas, but is also highly toxic so that such an animal would have quickly learned not to chew it. Note too that the mango pit is extremely hard, thus encouraging a large animal to swallow it along with the closely adhering pulp rather than try to chew it or spit it out. Consider also the papaya. The fruit are large and soft so that a large animal could easily take one into its mouth and just mash it lightly and swallow. Note too that the fruits of the papaya tree grow not high in the tree nor is the tree a low lying bush. Instead the tree is taller than a bush but its fruits are clustered at a height supermarket convenient for a large animal to pluck.

Barlow considers a number of other trees, the honey locust and the Osage orange, for example, as examples of ecological anachronisms, trees that have out-lived their mutualists and consequently must form new partnerships with other seed distributors or face extinction. For those trees that have pleased humans, the avocado, the mango, the papaya, etc., there is no immediate danger, but some other trees are at the edge of extinction. Their fruits fall to the ground and stay there until they rot. New trees grow only downhill when an occasional flood of water moves their fruit to a new location.

Barlow also sees ghosts from the Mesozoic era. She writes, "Ghosts of dinosaurs are easy to conjure in October and November wherever city landscapers planted ginkgo trees...even when I forget to look for the ghosts of dinosaurs my nose alerts me to their presence. Only a carrion eater could find the odor of fallen ginkgo fruit appealing. Before beginning this book, I wrongly blamed the alcoholic homeless for the vomitlike stench in Washington Square Park." (p. 12)

In short this book is about those trees—anachronisms—have been without their mutualists since the mass extinction of the mega fauna of the Western Hemisphere that took place about 13,000 years ago. It is a popular expansion on some original work done by ethnologist Daniel H. Janzen and paleontologist Paul S. Martin, their seminal paper appearing in the journal *Science* in 1982. Connie Barlow's prose is not only very readable, but is full of the excitement of scientific discovery, vivid and concrete, and packed with an amazing amount of information so that not only the trees, but the giant sloths, mastodons and mammoths—the ghosts of harvests past—come alive on the pages.

What Barlow does more than anything is open our eyes to the ecological nature of fruit and the relationships that exist between trees and the animals that eat the fruit. We learn how color, taste, aroma, texture, nutritional value, toughness of rind, size, shape, number of seeds and how they are encased, etc.—how all these qualities of fruit have evolved to entice the animals that will faithfully distribute the seeds, but also how some qualities discourage other animals, "pulp thieves" or "seed predators," that benefit from the food provided by the tree, but do not help in its propagation.

The story of the desert gourd was of particular interest to me because during many walks in the chaparral and deserts of California I have come across this vine with its hard, dry and unattractive gourds that were never picked or eaten. Barlow theorizes that the plant is also an anachronism, and that there did exist in the past animals that found the gourds, if not delicious, at least palatable.

Another curious anachronism reported on is the devil's claw of the Chihuahuan desert of Mexico. This plant produces a most amazing apparatus that wraps itself around an animal's foot and claw-like clings to the animal, dribbling its seeds to the ground as the animal moves. There is a photo of the claw on page 151 wrapped around a human ankle. Incidentally, the text is enhanced by a number of interesting black and white photos of the trees and their fruits.

This is one of the most interesting and original books on evolution that I have read in recent years, and one of the most informative.

Bearzi, Maddalena and **Craig B. Stanford** *Beautiful Minds: The Parallel Lives of Great Apes and Dolphins* (2008) *****
How dolphins and apes (and humans) are alike and different

Bearzi is the dolphin biologist and Stanford is the primatologist. The "parallels" between dolphins and great apes that the authors speak of are mostly in the use of what we call "intelligence" in their adaptations to life. Consequently this is a comparison of dolphin intelligence with primate intelligence, and of course implicitly with human intelligence.

The sections within the chapters are written first on one species and then on the other so that dolphin social behavior, for example, can be compared with primate social behavior (Chapter 4) or that their navigation through their differing environments can similarly be compared as in "Swimming with Dolphins, Swinging with Apes" (Chapter 3).

I was semi-surprised to learn that wild bottlenose dolphins even without hands have nonetheless been observed using tools—or at least one tool, a sponge worn as "a nose cap." The main speculation here is that the dolphins use the sponges "to protect themselves from a variety of harmful and toxic organisms near the sea floor and to avoid the abrasive sand, rocks, and broken shells that litter the deep waters..." (pp. 144-145).

Of course dolphins in sea shows have been taught to use balls and other objects as "tools for entertainment"—which brings me to this consideration: is dolphin intelligence limited by the fact that dolphins have no hands with which to use tools? The authors seem to think so, and at any rate the sponge use is the only example of tool use in the wild that they report, although the use of air bubbles to surround and confuse schooling fish can be seen as another bit of "tool" use, I guess. Which brings up the question of how much do we really know about dolphin behavior and intelligence? Observing animals in zoos or as part of a theatric show is one thing; observing animals in the wild is another. Animals in the wild behave in ways that may surprise us, and our knowledge of the use and extent of dolphin intelligence may be limited because we are not able to systemically follow them in the wild.

The same is true for chimpanzees and other primates. In the February, 2010 issue of "National Geographic" there is an interesting article by Joshua Foer (with photos by Ian Nichols) about an encounter with chimps in the Congo's Nouabale-Ndoki National Park. Unlike other places in Africa the chimps encountered here had apparently never seen humans before. Their behavior—full of curiosity and "approach/avoidance" displays including nesting overnight in the trees directly above the camp of biologists Dave Morgan and Crickette Sanz—proved most surprising. The chimps spent part of the night "testing" the humans by throwing down urine and feces onto the tents and howling! This is a bit different from other reports that I have read.

In "Beautiful Minds," the authors speculate on whether dolphins and apes have "a theory of mind"—that is, whether they are aware of what others may be thinking and whether they have a sense of self. Some dolphins were "marked with black ink in an area of their bodies not visible to them. They could, however, feel the ink. A mirror was offered, and the dolphins were watched to see if they were visually monitoring their bodies to find the ink spot." Some were. (pp. 180-181) This suggests self-awareness. In chimps it has long been known that they recognize themselves in mirrors and realize that the image in the mirror is not some other chimp. Here it is reported that an Asian elephant "was also able to pass the mirror self-recognition test, repeatedly touching a white X on the side of her head with her trunk." (p. 180)

What is clear to me is that the great intelligence demonstrated by chimps, bonobos and dolphins (and humans, by the way) is primarily the result of the need to understand and negotiate the complex social relationships they have with others. This is the key to the growth of these big brains. But intelligence defined as the ability to solve problems applies directly to the search for and procurement of food. Dolphins use cooperative hunting to surround and force to the surface schools of fish so that they cannot easily escape. As mentioned above, they even use bubbles to confuse and confine the fish. Apes use their minds to find and recall where and when they found fruits in season in a vast forest.

Ape intelligence is apparently limited by their inability to form abstract concepts, especially in terms of language that would allow them to pass on information to others. In the case of dolphins this is not so clear since we are at a loss when trying to understand what they are "saying" or why they do some of the things they do. It may be that we will find that dolphins do indeed have some sense of the abstract and can communicate about things such as fish not immediately present or actions and events in the past or imagined, which is the essence of human conceptual abilities. The trick is to have symbols such as words to stand for something not present or for acts not in evidence. Apes are limited in their ability to symbolize. Are dolphins so limited? We don't yet know.

Bearzi and Stanford in this very readable book have done a great job of bringing to a general readership some of the latest ideas and discoveries that are leading us toward a greater understanding of these unique beings, and of course to a better understand of ourselves.

Bloom, Howard *Global Brain: The Evolution of Mass Mind from the Big Bang to the 21st Century* (2000) *****
On the evolution of the planetary mind

Harold Bloom's *Global Brain* is one of those books, like Edward O. Wilson's *Consilience: The Unity of Knowledge* (1998), Jared Diamond's *Guns, Germs, and Steel: The Fates of Human Societies* (1997), and Ray Kurzweil's *The Age of Spiritual Machines: When Computers Exceed Human Intelligence* (1999), that presents the distillation of a lifetime of learning by an original and gifted intellect on the subject of who we are, where we came from, and where we might be going, and presents that knowledge to the reader in an exciting and readable fashion.

By the way, the very learned and articulate Howard Bloom (our author) is not to be confused with the also very learned and articulate literary critic Harold Bloom.

Bloom's theme is the unrecognized power of group selection, interspecies intelligence, and the dialectic dance down through the ages of what he calls "conformity enforcers" and "diversity generators." These diametrically opposed forces, he argues, actually function as the yin and yang of the body politic, active in all group phenomena from bacteria to street gangs. He is building on the idea that a "complex adaptive system," such as an ant colony or an animal's immune system is itself a collective intelligence. He extends that idea by arguing that a population, whether of humans or bacteria, is a collective intelligence as well. Put another way, intelligence manifests itself as an emergent property of a group. Furthermore, intelligence manifests itself as an emergent property of a collection of interacting groups.

This idea is certainly not original with Bloom—indeed it is part of the Zeitgeist of our age—but his delineation of it is the most compelling and thorough that I have read. It runs counter to the prevailing orthodoxy in evolutionary theory. In particular it is in opposition to Richard Dawkins's selfish gene theories and Ernst Mayr's insistence that natural selection operates on individuals not on populations. It is a synthesis of ideas that will, I believe, in the next decade or two, greatly alter the perspective of many of our scientific disciplines.

Bloom also posits "inner-judges" which function like biological super-egos; and "resource shifters" which function like neural nets, rewarding those strands of the group that are successful, punishing those that are not. To this he adds the playfully named "intergroup tournaments"; that is, war and other competitions between groups as close as human bands and as diverse as animals and their microbial parasites. Bloom defines these ideas on pages 42-44 and elaborates on them throughout the book with a summary in the final chapter.

The key idea that needs emphasis here is that Bloom believes (as I do) that evolution, cultural and biological, operates on groups as well as on individuals—groups of people, groups of animals, groups of microbes—cities, tribes, gangs, herds, species, bacterial colonies and viral masses. He sees all forms of life as interconnected in ways that are not obvious, but discernable if we find the right perspective. Bloom's perspective begins with the physics of the big bang, continues through pre-Cambrian microbial jungle, to the dialectic dance of Sparta and Athens, even to pre-September 11th Afghanistan (perspicaciously, by the way), until he concludes that all life on earth is, and has been, plunging toward an emergent property which might be called Gaia with a planetary brain.

Some observations:

"Reality is a mass hallucination" (p. 193) or "Reality is a Shared Hallucination" (title of Chapter 8; see also page 2 and page 170). This declaration, expressed somewhat differently, is a tenet of Buddhism, but here Bloom makes the case from a scientific point of view, and he makes it very well.

"Humans have been outfoxed...by a collective mind far older and nimbler than any we've developed to this point—the 3.5-billion-year-old global microbial brain." (p. 115) What Bloom is asserting here and throughout the book is that bacteria constitute a superorganism with an intelligence superior to ours that expresses itself through its complex chemistry and tactile behavior.

"...[T]he brain we think belongs solely to our kind achieves its goals by tapping the data banks of eagles, wheat, sheep, rodents, grasses, viruses, and lowly *E.*

coli." (p. 220) This dovetails with "We are modules of a planetary mind..." (p. 219) and "the global brain...is a multispecies thing" (p. 216), and the final line in the text, "We are neurons of this planet's interspecies mind." (p. 223)

In short, this is one heck of a book. And I'm just talking about the text, which is written in a spirited—sometimes even giddy—style that is infectious and thoroughly engaging. There are 66 pages of footnotes and a 62-page bibliography listing perhaps 500 titles. Some of footnotes contain multiple references, and of course there are errors. It is clear, for example, that human class did not exist 25 million years ago (as is asserted on page 148). When one looks at Bloom's footnote for the assertion, one realizes that he probably meant 25 *thousand* years ago. The point here is that we shouldn't be put off by all of his references. Those references allow us to check on his facts and gauge his interpretations. And, were any of us to actually read all of the approximately 500 titles he lists, I think we could at the very least apply for our own special ivory tower and some kind of honorary degree.

Bottom line: read this book.

Darwin, Charles The Voyage of the Beagle (1845) *****
An incredible adventure and a most enjoyable read

One of the amazing things about the voyage of the Beagle is that Darwin survived it! On the voyage south along the eastern coast of South America and then later on the western coast he would frequently take to the land and meet the Beagle at its next port of call further south or north. He would travel the land hiring gauchos or other guides and horses and mules so that he could study the geology and the flora and fauna. The hardships and dangers he encountered and survived would in some ways put Indiana Jones to shame. In Patagonia amidst the constant gaucho and Indian wars, rife with wanton bloodshed and a kind of genocidal determinism, Darwin rode on horseback and slept on the ground and ate mostly animal flesh of all kinds, including mare's flesh. In Tierra del Fuego the cold and barren lands were enormously forbidding, the inhabitants savage and the dangers very real. One senses in the young Charles Darwin a determination to be the kind of naturalist who leaves no stone unturned, no ridge unclimbed and no species uncollected.

What most surprised me was how well and vibrantly he described the many people he met. Here he speaks of the governor of St. Fe: his "favourite occupation is hunting Indians: a short time since he slaughtered forty-eight, and sold the children at the rate of three or four pounds apiece" (from the entry of Oct 3 and 4, 1832). And here is his description of Queen Pomarre of Tahiti: "The queen is a large awkward woman, without any beauty, grace or dignity. She has only one royal attribute: a perfect immovability of expression under all circumstances" (entry of November 25, 1835). Darwin was quite taken with the Tahitians lauding their sobriety (thanks to the temperance movement of the missionaries) while at the same time bringing a flask of spirits on his travels there. He seemed unaware of any inconsistency.

I was also surprised by Darwin's vigor. I had thought that he was prone to being sickly, and indeed at times, he reports that he was confined to his quarters and that he suffered from seasickness and even homesickness; but when one considers all the miles he travelled on foot, on horseback, and all the mountain peaks he obtained, and the deserts he crossed, the many insects bites he endured, and the hard, cold and wet ground on which he often slept, one has to applaud his strength of body and character. Another surprise was the amount of time he devoted to geology and speculations about the how the land came to be the way he found it. When he spoke of how the land had risen and the mountains formed I had the sense of how thrilled he would have been to have had the modern understanding of plate tectonics.

At a couple of points in the narrative, Darwin speaks of how the most luxurious vegetation does not support the greatest number of animals, or the largest. He compares the plains of Africa and Patagonia with the Brazilian rainforest and speculates on why this should be. At no point does he use the term "grasslands," and so I think we can conclude that he didn't have the knowledge we have today about how fertile grasslands can be, nor did he realize that most of the nutrients in the rain forest are contained within the living plants and organisms above ground leaving the soil relatively poor compared to grassland soil. In the entry for September 15, 1832, he writes: "In grassy plains unoccupied by the larger ruminating quadrupeds, it seems necessary to remove the superfluous vegetation by fire, so as to render the new year's growth serviceable."

Another bit of modern knowledge that would have pleased him to know is that the marine iguanas of the Galapagos Islands cannot just jump into the very cold water that exists there but must warm themselves first, and even then can only stand the water for a limited period of time (an hour or two, I believe). Darwin kept tossing one of the lizards into the water only to watch it return inexplicably again and again to the land.

I was looking for hints that Darwin was already thinking about natural selection, but the text contains nothing that I could find that is directly specific although at one point he refers to the origin of species as that "mystery of mysteries."

The book was written (and obviously rewritten and polished many times over) after Darwin returned to England after comparing notes with other naturalists. The advantage of this approach is the scientific rigor with which he is able to describe and evaluate his experiences. As a professional scientist, Darwin wanted to get all the scientific names right and avoid errors. One would expect through this approach that some immediacy would be lost, but if anything I suspect his journal gained in vividness and was made all the more intriguing for the precision of expression. It is, after all these years, still a most engaging and readable account of a most remarkable adventure—one of the best I've ever read, and I am surprised that it took me so many years to get to it!

The Voyage of the Beagle is also a book that will stay in print for many decades if not centuries to come, partly because it is so well written, and partly because Darwin is Darwin, but also because he *was* so precise in his descriptions of the animals and the people and the lands that he visited. By reading this we and future generations can learn of the changes that have taken place.

In short I was thoroughly dazzled at Darwin's enormously wide range of knowledge. But I shouldn't have been. In just reading this journal, one can easily see that young Mr. Darwin was already a superb naturalist and a brilliant thinker and observer.

Dawkins, Richard *The Ancestor's Tale: A Pilgrimage to the Dawn of Evolution* (2004) *****
A tour de force by a great teacher

The conceit here is to imagine that we can go backward in time to when our linage split off from some other branch, say to when the evolutionary line that has led to *Homo sapiens* split off from the line that has led to chimpanzees. That would be Dawkins's Rendezvous point #1 on page 100 which took place about six million years ago.

Like the characters in Chaucer's "Canterbury Tales," we are pilgrims traveling toward a spot, not in geography, but in time; and along the way we meet up with other pilgrims, eventually with all of life's creatures until we are gathered together in Rendezvous #39 where we meet our most distant relatives, the eubacteria. At each rendezvous point there is a "concestor," the common ancestor of all the lines of creatures that are now joined. Thus Concestor 1 was the "250,000-greats-grandparent" of both humans and chimpanzees. What Dawkins does at each point is to imagine what the concestor might have been like. Thus Concestor 1 would be an ape probably more like a chimpanzee than a human, a hairy ape that probably didn't walk upright, etc.

Prior to Rendezvous 1 there is Rendezvous 0 in which all of humankind meets together with the humans that went out of Africa 30,000(?) years ago.

At each of the rendezvous points there are tales, as in Chaucer's work. There is "The Tasmanian's Tale" in which it is related how the Tasmanian people were slaughtered. There is "Eve's Tale" in which we learn about mitochondrial DNA which is passed exclusively from mother to daughter. And later there are tales from the duckbilled platypus, the lungfish, the grasshopper, the redwood tree, etc. In each of these tales Dawkins, in his inimitable manner, focuses on some aspect of evolutionary biology and illuminates it for the reader. Some of the tales are difficult, but most are easy to read and fascinating to follow.

One of the strengths of this book is in the clear and engaging way that Dawkins presents the material and his forthright expression, and his honest way of separating what is his informed opinion from that of a scientific consensus. One thing I have always admired about Dawkins is his ability to be candid about what he knows and what he doesn't, and his courage in expressing unpopular opinions.

The question is, should the general reader buy this book and invest the time and effort in following Dawkins's pilgrimage? For those familiar with the great evolutionary biologist's work, this question answers itself. A new book by Richard Dawkins is an event to be celebrated. For others who want to learn something about evolution from one of its foremost experts (and celebrities, by the way) this book is highly recommended. The sheer expanse of Mr. Dawkins's knowledge and his infectious enthusiasm will inform and delight the reader. I suspect that most readers will learn more about how evolution works from this single book than from almost any other that I have read, and that includes works by such illuminati as Stephen Jay Gould, Ernst Mayr, and Edward O. Wilson. Incidentally, I believe that *The Ancestor's Tale* is a better read for most people than Gould's monumental *The Structure of Evolutionary Theo-*

ry (2002) if only because it is more accessible and less technical—not to mention about 750 pages shorter!

Some quibbles: I think Dawkins overrates sexual selection in bringing about evolutionary change in some cases, e.g., in bipedalism, and in the growth of the human brain (cf. Geoffrey Miller in *The Mating Mind* (2000)). It seems to me that natural selection had to be the driving force in something so felicitous as freeing our arms to carry and to work, and in the development of something so expensive and radical as a greatly oversized brain.

I also think that his idea to identify the lake, for example, created by the beaver, as part of the extended phenotype of the beaver is going a bit too far. I prefer to think of the lake as part of the beaver's "culture," and the behavior that created the lake as part of the phenotype of the beaver. Our airplanes and computers are part of our culture. Our ability to make them is part of our phenotype. (For Dawkins's full expression on this topic see *The Extended Phenotype* (1982).)

Here are a couple of examples of Dawkins's entertaining and elegant style:

Referring to the fall of the dinosaurs, he facetiously imagines that Shelley's celebrated poem, "Ozymandias," which mocked the vainglorious works of kings, was entitled "Ode to a Dinosaur." Indeed, how the mighty have fallen. (p. 255)

Talking about a sea squirt larva which gives up its mobile life to become sedentary (with attendant changes in neurological form), Dawkins remarks, "...when the time comes, [it] settles down to a sedentary life and 'eats its brain, like an associate professor getting tenure.'" (p. 370)

One of the best parts of the book, not to be missed, is the concluding chapter, "The Host's Return" in which Dawkins talks about re-running evolution. He speculates on which adaptations might and might not be returned. Certainly flight, which has evolved independently some 40 to 50 times would be seen again. But would the "swollen" human brain and the development of syntactic language evolve again?

Although Dawkins has addressed the so-called argument from design many times, I don't think he has ever demonstrated its paucity more eloquently and succinctly than he does beginning on page 549 where he calls it the "Argument from Personal Incredulity." His point is that to say something must have been designed because it is so complex really says "less about nature than about the poverty...[of the speaker's] imagination." On page 602 he adds, "Ultimately design cannot explain anything because there is an inevitable regression to the problem of the origin of the designer."

And finally I have a question for Professor Dawkins: why are there no viruses among the pilgrims?

Dawkins, Richard *A Devil's Chaplain: Reflections on Hope, Lies, Science, and Love* (2003) ****
A revealing collection of essays by a passionate scientist

One of the wonderful things about this book is the sense that one gets of a distinguished scientist letting his hair down, as it were, and discoursing informally on a number of interesting subjects including some outside his area of expertise. In the game of "Who would you invite to dinner if you could choose anybody?" Oxford University Professor Richard Dawkins, author of *The Selfish Gene,* and other important works on evolution, would be near the top of my list.

Not that I agree with everything he says. Indeed, that is part of the fun. Dawkins is adamant on some subjects, religion being one of them. A goodly portion of this book is devoted to letting us know exactly how he feels about the "God hypothesis," "liberal agnostics," and the so-called miracles recognized by especially the Catholic Church. The title of Chapter 3.3, "The Great Convergence" (of science and religion), for example, is used ironically. He sees no convergence; in fact, he calls such a notion "a shallow, empty, hollow, spin-doctored sham." (p. 151)

Clearly Dawkins is not a man to mince words. But his insistence on a restrictive definition of "God" as "a hypothetical being who answers prayers; intervenes to save cancer patients...forgives sin," etc., is really the problem. He considers the "religion" attributed to scientists like Einstein, Carl Sagan, Paul Davies and others (and even himself!) to involve a misuse of the term, calling such a definition "flabbily elastic" and not religion as experienced by "the ordinary person in the pew." (p. 147)

But what Dawkins is really railing against is the illegitimacy of believing in the supernatural and science at the same time.

While I think Dawkins makes a good point with this argument, I think it would be better to make a distinction between fundamentalist religion, which has been, and continues to be, the root cause of much of the horror in the world, and the more progressive varieties which recognize the limitations of the barbaric "Bronze-Age God of Battles." See Chapter 3.5 "Time to Stand Up" in which Dawkins rightly condemns the hatreds and violent history of the three middle eastern religions. At the same time I think he needs to realize that it is legitimate to define "God" as God is defined in, for example, the Vedas; that is, as The Ineffable, which has no attributes, about which nothing can be said.

However it is exactly his point that there is no evidence for the God hypothesis and that to partially accept such a notion, or even to be "agnostic" is to depart from a purely scientific viewpoint. In this I think the atheistic Dawkins is mistaken. Absence of proof is not proof of absence, period. And as far as religion, per se, goes, I would add that not only is religion part of human culture (for better or for worse), but is also part of the so-called "extended phenotype" of human beings, and not something that is going to be argued away.

I also have some reservations about his reasons for not debating with creationists. He believes that to debate with them gives them a legitimacy they don't deserve. In Chapter 5.5, he reveals a letter he wrote to Steven Jay Gould expressing such a view. I don't debate creationists either, but my reason is that creationists don't really debate. They have already made up their minds and are not capable of being influenced by evidence. Theirs is purely an exercise in propaganda. Furthermore, as Dawkins discovered himself (in Chapter 2.3 on the Australian film crew that he allowed into his house for an interview), it is often the case that creationists don't play fair.

In Chapter 1.5 "Trial by Jury" Dawkins presents his reservations about "one of the most conspicuously bad good ideas anyone ever had." I understand his demurral, but would like to point out that juries dispense a social justice; that the tribe makes its decisions based on what it perceives as good for the tribe now, not necessarily what's true in an objective or scientific sense.

Interesting enough, Dawkins demonstrates his knowledge of other scientific subjects, including physics, and he does it very well. I was particularly impressed with his explanation of entropy and how it effects the evolutionary process in Chapter 2.2. (See especially page 85.) He also does a fine job of elucidating why Lamarckism cannot work without a "Darwinian underpinning" since there must be a mechanism for selecting between the acquired characteristics that are improvements and those that are not. (p. 90) Good too is his characterization of genes as constituting "a kind of description of the ancestral environments through which those genes have survived." (p. 113)

On his tiff with Gould, Dawkins attempts to make amends by reprinting some semi-gracious and mostly positive reviews of some of Gould's books; however it is obvious that his professional and emotional differences with Gould remain.

One of the most important points that Dawkins reaffirms here is his belief that we humans, because of our unique insight into ourselves and our predicament, "can rebel against the tyranny of the selfish replicators." (p. 11) What Dawkins means is that we do not have to take biology as destiny or to take Darwinism as a template for our morality—a point often missed by his critics.

There is much, much more of interest in this refreshingly personal collection of essays by one of our most original evolutionary thinkers, some of it first rate, and some of it rather ordinary; yet taken in total reveals a lot about Richard Dawkins, scientist, science writer, teacher, and human being that I was pleased to learn.

Incidentally, the title is from Charles Darwin who speculated on how such a personage might regard "the clumsy, wasteful, blundering low and horridly cruel works of nature." (p. 8)

That "devil's chaplain" here is Richard Dawkins himself who mostly directs his ire toward the stupidities of human beings.

Dawkins, Richard *The Extended Phenotype: The Long Reach of the Gene*
(1982;1999) *****
Difficult but eminently worthwhile

This is a long and difficult book, although not as long and difficult as it might be if it had been written by somebody without Richard Dawkins' gift for clarity of thought and expression.

The crux of Dawkins' thesis is expressed early on and much of what follows is a very detailed supporting argument. What he wants us to see is that the "self-ish gene" has a reach that extends beyond the confines of the individual or-ganism that houses the gene. The phenotype of our genes is the human organ-ism in all its glory; however the extended phenotype of our genes is not only the human organism but part of the environment in which the organism finds itself. In other words, the gene has the power to influence not only our behav-ior but the behavior and structure of elements in the world in which we live.

This thesis is not as striking to me as it has been to many others mainly be-cause I have studied Eastern religious views, and it is a tenant of such views that the distinction between ourselves (the "selfish organism," in Dawkins' terminology) and the environment is an artificial one, an illusion actually. We are part and parcel of all that is around us and within us, and the boundary of our skin is merely functional. We cannot be understood by looking at only our bodies. Dawkins makes the point that looking at a beaver and microscopically examining it and its genes is not sufficient to an understanding of what a bea-ver is. We have to also consider the dams that the beaver builds, the trees that it gnaws down and even the streams that it dams and turns into lakes.

Presenting a point of view somewhat at odds with that of Dawkins (and one that I think that Dawkins does not sufficiently appreciate) is Franklin M. Harold in his book, *The Way of the Cell: Molecules, Organisms and the Order of Life* (2001). He writes, "Organisms process matter and energy as well as infor-mation; each represents a dynamic node in a whirlpool of several currents, and self-reproduction is a property of the collective, not of genes.... DNA is a peculiar sort of software that can only be correctly interpreted by its own unique hardware.... [S]ending aliens the genome of a cat is no substitute for sending the cat itself—complete with mice." (p. 221)

Dawkins tries to discount the view of those he calls "group selectionists" who see life from a "group benefit" viewpoint. Dawkins has, since writing this book, stepped back from this position to allow that some group selection may take place. I believe some day he may see the world not from a "selfish gene" point of view, and not from a "selfish organism" point of view, but from a "selfish ecosystem" perspective—well, more likely his successors will see this, since the work of a lifetime is not easily amended in one's later years.

Dawkins gives what he calls "our own 'central theorem' of the extended phenotype" on page 233: "An animal's behaviour tends to maximize the survival of the genes 'for' that behaviour, whether or not those genes happen to be in the body of the particular animal performing it."

This is a mouthful. Clearly we can say that the genes of the reed warbler code for behavior that benefits the genes of the cuckoo who has laid its egg in the warbler's nest. This is what Dawkins has in mind. But then arises the question, "how far afield can the phenotype extend?" Here Dawkins gets cautious and writes, "The farthest action at a distance I can think of is a matter of several miles." (p. 233) Note the chosen terminology, "action at a distance." This is from physics of course causing Dawkins to ask if there is "a sharp cut-off" of the genes' reach or "an inverse square law" at work?

It is here that I believe Dawkins has come so, so close to that which he will not see (or couldn't see then), namely that everything works toward an ecology and that the idea of selfish genes and selfish organisms is a limited view. In truth the reach of the genes should be governed by something like an inverse square law since humans are now reaching beyond the solar system.

When we look at such great distances we might want to credit the dreaded and verboten "group selection" that Dawkins is at pains to reject. Just as some see our earth as "Gaia," an organism itself, so too might we see those organisms that have the means to survive the destruction of the home planet by migrating to other planets as being selected by group as opposed to other groups who have no such ability. Planet A produces beings that extend beyond their solar system; planet B produces beings that do not. Both planets blow up. Who is "selected" by the (extended) environment and who is not?

Dawkins is one of the geniuses of science, and I don't mean to argue with the great insights he has brought to biology, but my point is that it is always something of an artificiality to speak of living systems as confined to one level of existence or expression. We may think of earth creatures as being completely separate from the rest of the universe, yet without the sun, 93 million miles away, we would not exist; and come a supernova even many light years away, we will be affected.

So all is one and one is all in some extended sense. And using the word "selfish" (as Dawkins knows) at any level of life is merely to be anthropomorphic.

Daniel Dennett, in a new afterword written in 1999, asks if this book is science or philosophy, and he answers both. I agree, and it is science and philosophy of the highest order, aimed equally at the professional and at the educated layperson.

Dawkins, Richard *The Selfish Gene* 30th Anniversary Edition (2006) *****
I think Dawkins is wrong in his central argument. Here's why:

The first thing I want to say is how much respect I have for Richard Dawkins as a scientist, as a teacher, as a writer of fascinating prose, and as a person. He is a brilliant and courageous man who works hard to bring his knowledge and insights to all of us.

The second thing I want to say is that *The Selfish Gene* is one of the landmark science books of the 20th century, and so I am pleased to see this 30th Anniversary Edition (from 2006) with a new introduction by Dawkins and some new footnotes.

Rather than review the book as a whole, however, as has been done many times, in this review I want to concentrate on the central issue of the book, namely the question of "at what level does natural selection work?"

Dawkins believes that the environment selects certain genes, or more properly speaking, suites of genes and therefore operates primarily at the level of the gene. I disagree and believe this is like saying that the public selects certain letters, or words, or sentences of words when buying a book. The words (or more properly the ideas represented by the words) are the reason the public selects a book, but what the public selects is nonetheless the book. Genes are like ideas in books. Ideas must appear in some medium, even if it is just word of mouth. Genes must appear in organisms, which are the products of both the genetic instructions and the environment in which they develop. Consequently genes help to produce individuals (or in the case of social insects, a group of individuals that can be seen as a single organism). Dawkins calls these individuals "survival machines." In turn the environment selects certain survival machines that contain certain genes.

Another way of expressing this is to say that the environment selects genes by proxy, that is, through the medium of the individual phenotype. The environment cannot directly affect the genes since the genes are safely encapsulated within the survival machine which does not in any Lamarckian way communicate with them. The exception is when an electromagnetic particle hits the code and alters it, creating a mutation. The environment does not act on that altered code; instead it acts upon the individual that is born to carry that altered code or lack thereof.

The individual gene itself (if we can speak of such a thing which is just a section of code) doesn't work in isolation. It is always allied for better or for worse with other sections of code. Certain sections of code are reproduced again and again because they are handy or work well with other sections of code in a way that allows the survival machine to reproduce and its offspring to reproduce. But the environment cannot select certain selections of code. It can only select the individual containing that code (and a lot of other code besides). In fact, it cannot just select the individual, it must select its possible mates and even much of its environment as well, such as the plants and animals it uses for food and shelter. To speak of selecting genes or even individual organisms is just a convenient way of talking.

What is really selected is a group of organisms of some kind. Some consider an important group selected by the environment to be the species or the ecology. Giving a large enough perspective, I would go so far as to say (going beyond Lovecock and Gaia) that natural selection operates on the level of life itself.

Another point is that the genes never reproduce themselves by themselves. Nothing in this world that I know of actually reproduces itself by itself, except dividing cells, and they do this only most of the time. As is now known, occasionally bacteria trade genes with other bacteria and thereby reproduce not quite exact copies of themselves. A strand of DNA is replicated with the help of the machinery of the cell. Viruses need cells to replicate themselves. Anything that was one hundred percent effective in making exact copies of itself would not undergo Darwinian evolution and would in fact have died out long ago. The dreaded grey goo of nanobots replicating until they cover the earth is still just a fantasy of science fiction.

The problem with the current understanding of evolution and natural selection is the problem of not seeing that everything is connected. Any place we draw a boundary is artificial or arbitrary. Even at the skin. Franklin M. Harold, in his book, *The Way of the Cell: Molecules, Organisms and the Order of Life* (2001) writes, "Organisms process matter and energy as well as information; each represents a dynamic node in a whirlpool of several currents, and self-reproduction is a property of the collective, not of genes.... DNA is a peculiar sort of software, that can only be correctly interpreted by its own unique hardware....sending aliens the genome of a cat is no substitute for sending the cat itself—complete with mice." (op cit., p. 221)

For those of you who have read Dawkins' original edition from 1976, this edition is still to be recommended, particularly for the updated bibliography and for the 66 pages of endnotes where Dawkins graciously admits errors and points to new discoveries, most interestingly that of Zahavi's "handicap principle" which goes a long ways toward explaining some "altruistic" behavior. See my review of *The Handicap Principle: A Missing Piece of Darwin's Puzzle* (1997) by Amotz and Avishag Zahavi on page 199.

Gould, Stephen Jay *I Have Landed: The End of a Beginning In Natural History* (2002) *****
The tenth and final collection

I was a little bit disconcerted when I saw the title of this, Stephen Jay Gould's last collection of essays. I thought: has he anticipated his own sadly premature death with the metaphoric "I Have Landed" or is this a kind of melancholy coincidence, or perhaps I am reading into the title something different from what it warrants?

As it turns out, "I Have Landed" is not a reference to the Lethe shore of the poet, but a reference to his grandfather's arrival at Ellis Island on September 11, 1901, exactly, to the day, one century before the attack on the World Trade Center in New York. It is from this coincidence that Gould embarks upon some musings that form the touchstone for this, his tenth and last collection of essays.

He is a man who will be sorely missed, a complete original, at once the very embodiment of a meticulous scientist and an establishment New York liberal. He is one of our greatest essayists, a humanist and a quintessentially rational man who has often argued in favor of the value and importance of religious thought. Born in modest circumstance, descendent of Hungarian immigrants (as was another of our most prolific writers, Isaac Asimov) he fell in love (as he recounts in these pages) with the NYC Museum of National History as a child and never lost his love for "the odd little tidbits," nor his sense of himself as a natural historian. He is a "student of snails" (p. 324), a classical nerd "shorter than average" (p. 246) who spent more time at the Hayden Planetarium and the *Tyrannosaurus* exhibition than he did playing his beloved baseball, a pale-ontologist who became not only a gifted essayist but an international celebrity.

It's a neat trick what Stephen Jay Gould has done with his life, and it is a neat trick that he "chose" (if I may) to leave this vale of tears almost immediately after finishing not just this book, but more significantly, *The Structure of Evolutionary Theory*, the "life work" of his "mature years, twenty years in the making and 1,500 pages in the printing." It has been noted that people typically die after a long illness not the day before Christmas or the day before their birthday or the day before the christening of their youngest grandchild, but the day after. And the very great choose to leave us only after they have finished some compelling project to which they have devoted the last years of their life. Gould remarked in the Preface on the coincidence of his finishing these twin projects together in time for publication in the "palindromic" year of 2002—(how he loves the odd fact, the detail that others might miss, and how he rejoices in sharing such "tidbits")—while recalling the earlier "conjunction" of the near simultaneous publication of his first book of collected essays, *Ever Since Darwin*, and his "first technical book for professional colleagues," *Ontogeny and Phylogeny* in 1977. I wonder if he knew that these would be the bookends of his life.

This collection is touted on the blurb as "the most personal book he has ever published"; nonetheless it is very much like the nine other collections. There is the usual intricate and sometimes whimsical analysis of a bewildering range of subjects anchored to natural history with (of course) some asides on baseball. The style has gotten a trifle more ornate, the qualifications upon qualifications a bit more belabored, the subordinated clauses in the parallel construction of his architectured sentences a bit more in number, but otherwise he is still the same man, ponderously thorough and passionately alive in argument and analysis.

Some old subjects (the limitations of reduction in the biological sciences; the misleading popularizations of evolutionary ideas; the excessive ink the dinosaurs get, the delusion of racism, etc) are returned to and reworked. There is a convincing argument in favor of Vladimir Nabokov as a scientist in addition to his work as a literary artist. There is a return to Freud and his "evolutionary fantasy." (Freud could not shake himself from a Lamarckian view). There is a look into the origin and meaning and misuse of such words as "syphilis" and "evolution," noting respectively that science has done a poor job of treating syphilis and that the meaning of "evolution" has changed. (Darwin did not use the word in the first edition of *Origin of Species*, although, as Gould notes, he ended the book slyly with the word "evolved.")

Less anyone think that Gould is all learning and little insight (a laughable idea considering his contributions to evolutionary theory, his punctuated equilibria and his spandrels, to name the best known) consider this salient (and to some extent, self-addressed) question from page 4: "How do scientists and other researchers blast and bumble toward their complex mixture of conclusions (great factual discoveries of enduring worth mixed with unconscious social prejudices of astonishing transparency to later generations)?" How indeed do we escape the prejudices of our times, and to what extent does this apply to Gould himself?

There are drawings and black and white prints and in all 39 chapters in this handsome book. Gould ends where he began with his grandfather Papa Joe with thoughts about New York and its people adorned with the majestic cadences of Ecclesiastes, "To everything there is a season, and a time to every purpose under the heaven. A time to be born and a time to die: a time to plant, and a time to pluck up that which is planted...."

And a time to be landed. Gould, no doubt, has landed not on the Lethe shore but near the Cambrian sea where he might take a closer look at those myriad creatures to which he devoted so much of his life, and from which he learned so much that he was able to share with us.

Hallé, Francis, trans David Lee. *In Praise of Plants* (1999; 2002) *****
Technical but wondrously informative

It's always a good sign to see that someone has bothered to translate a science book from another language into English. Publishers can usually get some English-language scientist to write a tome on the latest discoveries in a more commercially agreeable manner than putting together a translation. So when the translation appears you know the book is good and/or original in a distinctive way.

In Praise of Plants by botany Professor Emeritus Francis Hallé of the University of Montpellier, France is such a book. However it is by no means a popular treatise; indeed, if you want to get the look and feel of a botany article in a professional journal, this book provides an entire book's worth! The material is technical, detailed, and uncompromisingly professional.

So why has the Timber Press chosen this volume to bring to the English speaking world? Partly because of the international prestige of Hallé, who is an expert on tropical plants; partly because they were able to get a translation by David Lee who is Professor of Biological Sciences at Florida International University; and partly because of the striking nature of Hallé's presentation.

Hallé emphasizes the form of plants and how that form has developed evolutionarily from their need to secure the services of both sun and earth while remaining nearly immobile. There are dozens of line drawings in the book, most by Hallé himself, illustrating the differences between plants and animals with the text explaining why these differences occur. For example, because plants are sessile (attached to the ground) they are symmetrical on the horizontal plane, a tree looking pretty much the same from whatever spot on the ground you view it. However in a vertical sense a plant is very different since its crown is in the air looking at the sun while its roots are in the ground looking for water and minerals. In contrast, animals (I'll just quote Hallé so you'll get a feel for the technical language): "have dorsiventral polarity and antero-posterior and bilateral symmetry." (p. 70)

Fortunately the attractive and sometimes funny drawings help to penetrate the language for this amateur!

Here are some examples of the sort of things you can learn from this book:

At the microscopic level, where gravity is relatively "negligible compared to other forces" like "surface tension, viscosity, friction and Brownian motion," (p. 64) life forms tend toward the round and take on the symmetries we associate with astronomical objects like the sun and Saturn. Hallé gives examples of bacteria, amoebas, diatoms, etc. where "vertical polarity simply does not exist." (p. 64) Science fiction writers take note: creatures living in interstellar dust clouds will be more or less round.

One of the clear homologies (same form) assumed by plants and animals is in "the external (assimilating) surface of a plant and the internal (digestive) surface of an animal." (p. 51) The plant maximizes its surface area to expose as much of it as possible to the sun and the air, while the animal creates folds and such within its alimentary canal so as to provide a large surface area for effective digestion. Hallé notes that plants resemble fractals externally. (p. 52)

The waste products of animals bring forth (to our sensitivities) malodorous compounds as do their decomposing bodies. Hallé explains why this is so on pages 148-151, and why the waste products of plants and their decomposing bodies do not usually offend us; indeed the smell of new mown hay and forest humus or even a compost pile, can be very agreeable. On page 149 he favors us with a drawing of a tree which grows in part upon the waste products of its metabolism stored in its trunk. Next to the tree Hallé has a dog on top of a pile of its excrement, noting that "An animal that stored its excrement would also be capable of becoming very tall."

Hallé's love of plants and his deep respect for them, and his life-long experience in studying them comes through most wonderfully in this fine book. Although technical, it is accessible to amateur botanists and just plain old gardeners and lovers of plants with just a little effort.

Jones, Steve *Darwin's Ghost: The Origin of Species Updated* (1999) *****
Darwin's argument, much advanced

Darwin had his bulldog in Thomas Henry Huxley, and perhaps his pit bull in Richard Dawkins, and now he has his ghost in the person of Steve Jones who avers that Darwin's "spirit is on every page" of this eminently readable book. It would be hard to argue with that since the chapter by chapter plan of Jones's book closely follows Darwin's and many of the examples of evolution at work are elaborations on Darwinian themes. As Jones tells us in the Historical Sketch that begins his book, what Darwin was at pains to accomplish in *The Origin of Species* (1859) was to make "a bold statement of the idea of evolution" while at the same time produce "a work of persuasion as to how it took place."

Darwin knew that the persuasion would be the hard part. Now more than a hundred and fifty years later, many people are still not persuaded, perhaps the vast majority of people. In his introduction Jones notes that, according to an opinion poll taken in 1991, a hundred million Americans believe that "God created man pretty much in his present form at one time during the last ten thousand years." So Jones too has his work cut out for him. Although his lively prose is perhaps more accessible to a modern reader than Darwin's Victorian cadences, he, like Darwin, will reach only a very small minority of the human race. That's a shame since Jones's arguments and evidence for the veritable fact of evolution are overwhelmingly powerful and impossible to deny. They are also fascinating to read. Some examples:

So powerful is the process of evolution that zoos, human institutions that attempt to preserve threatened species, cannot. Instead the animals evolve within the constraints of their new environment and become (eventually) altered versions of their wild ancestors (p. 36).

On the differentiation of sex cells into sperm and egg, Jones writes, "Long ago...sex cells were all the same size and fused to make an embryo... Then self-interest made an appearance and one partner moved to making smaller but more abundant cells. He (for such was, from that moment, his gender) might have hungry young, but there were more of them" (pp. 81-82)

On the "Cambrian Explosion": "...a failure of the geological record rather than of the Darwinian machine. Its radical new groups reflect not a set of exceptional events, but something more banal: the first appearance of animals with parts capable of preservation" (p. 207)

Professor Jones does not limit himself strictly to observations on evolution. His erudition includes references and allusions to literature, classic and modern, notably Shakespearean, where the grave digger from Hamlet makes an appearance in order to further our knowledge of the decomposition of buried bodies. Jones is particularly strong on using knowledge from other disciplines to illustrate the process of evolution. He notes, for example, that a new Hawaiian island, "to be named Loihi" is "under construction" and due to "break the surface in thirty thousand years" (p. 262) On page 287, we learn that there is a fresh water lake beneath the Antarctic ice that scientists want to drill into "in the hope of finding yet another universe of life." On page 231 we are reminded that four hundred million years ago our year was four hundred days long, the evidence coming from the growth rings of corals.

Part of the illuminating power of this book is in the effective use of metaphor and analogy. Thus a new island rising out of the ocean is compared to a new born child, waiting to be invaded by flora and fauna, grasses and/or bacteria, as the case may be. Or, on page 307, junk DNA is compared to "the letters in a word, still retained in the spelling, but become useless in the pronunciation." Twice Jones refers to species becoming nonsexual as "abandoning their males," an expression that sheds stark light on the nature of sexuality. Sometimes Jones decorates his text with sly, humorous asides, as on page 237 where he is discussing grape varieties he notes that "Britain has an Anything but Chardonnay club." Or on page 294 where he makes the observation that our brain has become "so elaborate as—so far—to be unable to understand itself."

The only weakness of this book—and perhaps it is not a weakness at all—is the conversational tone that contrasts somewhat with Darwin's laboriously cast sentences as he oh, so carefully advanced his argument. Jones knows that the argument is long past the point of being overwhelming. What is really needed is a greater acquaintance with the argument by a larger public. Jones's lively tome, packed with fascinating information, is a small, but welcome step in that direction.

Jones, Steve *Darwin's Island: The Galapagos in the Garden of England* (2009)
*****
Updating Darwin and his scientific interests

Steve Jones, who is a professor of genetics at University College London and a most engaging writer on evolutionary biology, wrote this book to coincide with the bicentennial of Darwin's birth and the hundred and fiftieth anniversary of the publication of "The Origin of Species." He calls his book "Darwin's Island" to emphasize the fact that the vast majority of Darwin's work was on the biota of the island of England following his return from the voyage of the Beagle and not on what he learned during the scant five weeks he spent in the Galapagos Islands as a young man.

Darwin wrote a four-volume work on barnacles (over a thousand pages); he wrote on "Orchids and Insects," on the "Expressions of Emotions," on the "Formation of Vegetable Mould by Earthworms," and of course on "The Descent of Man" and other works, comprising in total more than six million words. Jones' intent is to introduce the reader to the wider range of Darwin's work and by doing so demonstrate why Darwin is widely considered the greatest biologist who ever lived.

Jones' technique is to devote chapters to Darwin's many interests while bringing us up to date on the current understanding. Thus we read about what Darwin learned about worms, barnacles, insects, insectivore plants, sexual selection, our facial expressions, etc., and how that agrees with or differs from what modern science has discovered. What we find out is that Darwin was amazingly prescient in many areas mainly because he worked so diligently for so many years with the kind of enthusiasm few of us can muster. And it didn't hurt that he was a brilliant man.

Darwin could have been a man of leisure because of inherited wealth, but he was driven to discover as much as he could about the natural world. He immersed himself into scientific research, performing experiments as well as reading, and corresponding with other scientists and amateurs from around the world. He dug up the ground around Down House where he lived; he dissected specimens, he worried about the adaptive vigor of his children since he had married his cousin (hence his volume on "Cross and Self-Fertilisation"), he measured things, he explored the woods and streams and seashores of his English "archipelago"; he examined fossils, and all the while he pondered deeply on the nature of life and on how evolution works.

The effect of Jones' technique in showing both what Darwin knew in the 19th century and what we know today is to emphasize how the world has changed since Darwin's time. We learn how some species have circumnavigated the globe and caused other species to go extinct, especially how the "weediest" of all species, human beings, have altered and destroyed environments and brought about changes in our use of the natural world that would have probably appalled Darwin.

Being a geneticist, Jones knows very well what Darwin could only guess at, that is, how the traits of species are handed down, how "descent with modification" works. And that is another strength of this remarkable and very readable book, demonstrating as Jones does how much Darwin was able to understand and get right without any knowledge of the basic mechanism of inheritance as expressed in genetics. How he would have marveled at what we know today.

Jones closes by seeing a "triumphant of the average" as we and other weedy species scurry about the globe mating widely instead of closely as in Darwin's time when people and other creatures seldom encountered opportunities much distant from the place of their birth. He sees what I once called "the browning of society" as natural selection irons out the differences between equatorial humankind and those from northern climes, as Asians marry the English, as Russian tumble weeds spread across the American west. When once it was the rich who had the most children, today it is the poor. Jones notes that "The gulf has closed through restraint by the affluent rather than excess by the poor." He does not speculate on what this change will have on society, but posits that the opportunity for natural selection "is in steep decline," meaning I suppose that evolutionary change in humans will become increasingly static. Musing on how that will play out in the long run, Jones writes darkly: "For Homo sapiens, some nasty surprises no doubt lurk around the corner. Someday, evolution will take its revenge and we may fail in the struggle for existence against ourselves, the biggest ecological challenge of all." (p. 286)

Lavers, Chris *Why Elephants Have Big Ears: Understanding Patterns of Life on Earth* (2000) *****
Splendid and readable

Chris Lavers is a paleontologist who specializes in wildlife ecology. It is from this point of view that he presents some of the ideas and controversies of current evolutionary theory along with some of the excitement of recent discoveries and understandings in a popular and nontechnical manner. His readable text is aimed precisely at the educated non-specialist, but without a hint of any dumbing down.

In the title chapter we learn that elephants pump the warm blood from the interior of their bodies to the array of tubes in their ears to dissipate excess body heat. From this consideration Lavers is led to a discussion of whether dinosaurs were warm blooded or not. The evidence he presents makes it clear to this observer that they were, but his cautious conclusion is that the case hasn't been proven quite yet. Lavers hints that the dinosaurs may have to be put in another category, perhaps somewhere between warm blooded and cold, or maybe even somewhere beyond. How about: "I'm hot-blooded, check it and see" (to reprise a rock lyric).

Lavers goes to considerable depth to demonstrate how much we can learn by combining evidence from the fossil record with what we know about the metabolism of animals and how their bodies work. Dinosaur anatomy, for example, strongly suggests a closer kinship with today's avian world than with the reptilian. Furthermore, the large size of many dinosaurs is inconsistent with cold-bloodedness. Reptiles can't get as big as a Brontosaurus because (for one thing) they would not be able to regulate their temperature. Lavers points out that all the really big animals on earth today, with the exception of the giant tortoises, Komodo dragons and some snakes—and they aren't really that big— are warm-blooded. He cites the arguments of Robert Bakker and others to conclude that *T. Rex*, for example, wouldn't have the metabolic power to run down prey if it were cold-blooded.

I found Lavers's discussion of the difference between non-oxygen-based metabolic reactions capable of "supercharged" bursts of short-lived energy typical of reptiles, and the sustainable aerobic reactions typical of mammals like dogs and humans very interesting. The quick bursts are those of the sprinter who is wasted after at most a few hundred yards, while the aerobic engine sustains the pace of the long distance runner. Also interesting is the material in the chapter "Life on the Edge" about how birds and mammals maintain their body temperatures in the climate extremes of the deserts and the polar regions of the earth. Lavers notes that in very cold places there are no reptiles.

In some of this I am reminded of the famous and splendid essay by J. B. S. Haldane, "On Being the Right Size," published many decades ago. Lavers presents the same kind of reasoned argument based on physiology and anatomy to demonstrate why animals are built the way they are and why it would be difficult for them to be constructed otherwise. One comes away from the reading with a sense of having learned something important and exciting, a sense of having acquired understanding, not merely a collection of facts.

Mayr, Ernst *What Evolution Is* (2001) *****
A splendid first course in evolution

(But not for dummies.)

Perhaps the most remarkable thing about this book is that the author was born in 1905. What legendary biologist Ernst Mayr might next want to share with us is his secret for remaining so mentally acute for so many years! Reading this exposition on evolution by "The world's greatest living biologist and a writer of extraordinary insight and clarity" (Stephen Jay Gould, on the jacket cover) is somewhat like taking Evolution 101 as it might be taught by Professor Mayr. As he writes in the Preface, his purpose is didactic. He would like us to know more about evolution and how it works.

First he presents the evidence for evolution, explaining (I hope) once and for all how evolution can be established as a fact even though we cannot perform experiments as we might in physics or chemistry: "Evolution...must be inferred from observations. Such inferences subsequently must be tested again and again against new observations, and the original inference is either falsified or considerably strengthened..." (p. 13) He adds on page 276, "I cannot see why...an overwhelming number of well-substantiated inferences is not scientifically as convincing as direct observations. Many theories in other historical sciences, such as geology and cosmology, are also based on inferences. The endeavor of certain philosophers to construct a fundamental difference between the two kinds of evidence strikes me as misleading."

To this I might add that *all* the evidence we have of the external world is from inference. Even so-called direct observations (whatever they may be) are inferences from the evidence of our senses and must be checked against the same inferences that others make.

Next Mayr explains how change and adaptation take place. He then explains why there is biodiversity. These are the first three parts of the book. Part Four is on human evolution and Chapter 11 in particular is a splendid, concise interpretation of the evidence for human evolution.

One of the thorny issues Mayr addresses is selection. He explains that it is the individual (the phenotype) that is selected, and not the gene and not the population. "[A] gene as such can never be the object of selection" because it "is only part of a genotype, whereas the phenotype of the individual as a whole (based on the genotype) is the actual object of selection." (p. 126) The gene cannot be the object of selection for another reason, namely that a single gene seldom, if ever, acts independently of other genes. They work together to bring about some feature of the phenotype and are subject to the action of regulatory genes (hox and pax genes). (p. 127) Furthermore, "Many genes do not have standard selective value. A gene may be beneficial when placed in one particular genotype, but it may be deleterious when placed in a genotype with different genes." (p. 128)

One of the things I learned here (p. 129) is that the phenotype includes "all the products of the behavioral genes. This includes the nest a bird builds, or the web of a spider, or the path of migration of a migratory bird." It also includes the gametes. Thus the ability of a spermatozoon to "swim" is part of the phenotype and is subject to natural selection.

Another interesting issue is group selection. Mayr defines two group types, "casual groups" and "cohesive social groups." Members of the former "are associated in a group [that] makes no contribution to their fitness." The latter, however, "owing to social cooperation among its members" "can indeed be a target of selection." This cleared up the group selection fuzziness for me.

It is interesting to note, however, that Mayr's argument seems to imply that if the cooperating group is the same as the species, then a species can be selected. However he writes on page 280, "The species as a whole is never the target of selection." He explains that "the differential success of [an] entire species is superimposed on...individual selection." Or, if I may phrase it another way, the differential success of a species is the result of the differential success of its individual members. What this really means, however one wishes to phrase it, is that selection can apply to an entire species (through its members).

A very fine example of Mayr's intelligence and sensitivity can be gleaned from reading his answer to the question on page 262, "Are there human races?" There are indeed races, Mayr explains, but the "race problem" is a result of "a faulty understanding of race. These people," he continues, "are typologists, and for them every member of a race has all the actual and imaginary characteristics of that race. To translate this bias into an absurd example, they would assume that every African-American can run the 100-meter dash faster than any European-American." What a race is, is a population and its members are individuals, not types. This is true of species as well.

There are a number of other technical and crucial issues in evolution that Mayr addresses including saltation and punctuated equilibria, altruism, kin selection, speciation, the origin of birds, etc. He even goes into a little exobiology on page 263. The book includes two appendices designed to help the reader cope with criticisms and questions about evolution. Appendix B sets forth 24 questions about evolution, such as "Is evolution a fact?" (yes) and "Is the Gaia hypothesis incompatible with Darwinism?" (no), etc. There is a glossary and an excellent index. There is some repetition, but I think we can take that as emphasis since this is an exercise in public education.

Although Mayr uses a minimum of jargon and writes in a straightforward manner, the issues are not simple. They need to be studied to be understood and appreciated. This is why I call this book Evolution 101 by Professor Mayr.

Morris, Richard *The Evolutionists: The Struggle for Darwin's Soul* (2001) ***
Cat fighting among the old guard in evolutionary biology

Well, Darwin's soul really isn't up for grabs. What is at stake is just who among the illuminati of the Darwinian establishment really have the goods on how evolution works and how it doesn't. "Spandrels" of the mind, "habitat tracking," how complexity affects evolution, "species sorting," whether evolution proceeds by leaps and bounds or just plods along, and other contentious matters form the body of this unsteady but interesting book.

The main antagonists are the usual suspects, Stephen Jay Gould, Niles Eldredge, Richard Lewontin, et al., proponents of punctuated equilibrium and a "holistic" approach to evolution on the one side, and Richard Dawkins, Daniel Dennet, John Maynard Smith, et al., gradualist reductionists, the so-called "orthodox Darwinians," on the other. Dawkins, et al. believe that natural selection is the only really important factor in evolution while Gould, et al. believe that natural selection alone cannot fully explain how evolution works. Morris reviews their various publications and quotes them, revealing that they behave rather badly at times, sometimes resorting to unseemly personal attacks on one another—which leads me to observe that Darwin, who never involved himself in hot debates, much less in name calling, must be turning over in his grave.

The irony is, as Morris fumbles to makes clear, the seemingly substantive differences that are being so hotly debated are for the most part actually ones of emphasis and interpretation. Nobody involved doubts the supremacy of natural selection as the driving force in evolutionary change, any more than any of them doubt the fact of evolution. Morris gives the reader some background information about evolution and introduces complexity theory in order that the debate may be followed. In the penultimate chapter he gives a summary of the evidence as he sees it. A final chapter entitled, "Controversy and Discovery," includes the currently hot idea "that evolution can proceed at a more rapid rate than anyone had suspected." (p. 233) There is an annotated bibliography and a useful appendix listing relevant Web sites. Morris tries to avoid taking sides in this debate. Indeed, he bends over backwards to be fair, and that attitude, along with a beguiling, easy to read style, is the strength of the book.

There are weaknesses, however. His focus is too narrow with its concentration on Gould, et al. and Dawkins, et al. and their differences when there are much more interesting and immediate questions currently being debated. (I imagine that the young lions in evolutionary biology are very tired of seeing those old guys still getting all that ink!) For an interesting book by a young evolutionary psychologist on some of the newer controversies see Geoffery Miller's *The*

Mating Mind: How Sexual Choice Shaped the Evolution of Human Nature (2000).

And then there are all those typos! I found typos on pages 41, 47, 107, 114, 199, 203, 228, and 232. In one case the word "would" was left out. In another the word "out" was used when the word "at" was meant, and in a couple of places extraneous words were left in. For example, on page 203 a sentence begins, "You should not should not automatically conclude..." When one sees a lot of typos in a book it suggests that the author did not read the proofs, or if he did, he did a cursory job of it.

Worse than the typos (and if I found eight, there are surely others) are some misstatements of fact and intent. On page 34 he writes that the mammals that survived the K-T extinction "are not more <evolved> than their dinosaur predecessors." As Morris points out on page 32 "a frog is just as <evolved> as a human being." But that means frogs living today and human beings living today. To compare how "evolved" the dinosaurs living 65 million years ago are to mammals living today makes little sense. Note too that on page 34 Morris refers to the extinction of the dinosaurs as taking place 65 million years ago, which of course is the standard take, but on page 124 he unaccountably states that the "collision with an asteroid" took place 70 million years ago. Actually he writes, "70 millions years ago," which, I just noticed, is another typo!

There is also entirely too much repetition in the book, as though the chapters were independently conceived and meant to be published separately and then not properly edited. For example on page 204 Morris repeats the same ideas, and even some of the same wording, that appears on page 123. Chapter 8, "The Evidence," in particular contains a lot of unnecessary repetition.

Finally there is a most annoying error on page 175 in Morris's discussion of the Watson selection task. As written the instructions are incomplete and must leave readers scratching their heads about what is given as the correct answer. He writes:

"Suppose you are shown the four cards marked with the following symbols: D F 3 7 You are then asked which two cards you must turn over to see if any of the cards violate the following rule:

If the letter D is on one side, then there will be a numeral 3 on the other.

Which two cards do you turn over?"

Morris's answer, cards, D and 7 is partially correct, but what about the card with the F? According to the directions it also has to be turned over (to see if there's a D there) making it three cards that must be turned over, not two. This error resulted because Morris left out the following proviso, namely that the cards always have a letter on one side and a number on the other.

This is an excellent idea for a book, but I don't think Richard Morris realized its potential.

Ruse, Michael *Darwinism and Its Discontents* (2006) ****
Adjudicating points of contention

In a previous book, *Can a Darwinian Be a Christian?* (2001), Michael Ruse reconciled the ways of Christianity to the fact of evolution. He did so by allowing for a largely symbolic reading of the Bible and by defining Christianity as a system of belief about matters beyond the reach of Darwinism, such as our having souls and being made in the image of God and being given heavenly or hellish eternal life in realms not subject to biological evolution.

Here Ruse adjudicates various disputes between Darwinism and its critics and among Darwinians themselves on such matters as natural selection (especially this), punctuated equilibrium, group selection, drift, reductionism, etc. Unfortunately I don't think he adds much that is new to the discussion, and his torturously "correct" navigation between believers and non-believers left this reader annoyed. Spill the beans! For example, state it clearly: Christianity that relies on a literal interpretation of the Bible is incompatible with Darwinism. Period. Add: Those who appreciate the fact of biological evolution cannot accept that man was made in a Christian God's image or that a personal God is, and has been, shaping events on this planet.

Ruse writes from the point of view of a historian of evolutionary science and as someone sympathetic with what I might call progressive Christianity, a Christianity that knows that the world was not made in six days and that the earth has been around for a few billion years and that God does not have a belly button or even an alimentary canal or a need for either. Ruse is an expert on Darwinism and its contemptuous history and he understands the major issues very well. He is the kind of writer who bends over backwards to be fair to his opposition, such as creationists and Intelligent Designers, as well as atheists like Richard Dawkins or agnostics like Stephen Jay Gould. He is also the kind of writer who equivocates a lot, whose instincts are to find common ground and to further responsible and honorable dialogue, which is the strength of this book.

He begins with a chapter demonstrating that Charles Darwin really does deserve the credit he gets for being the first to understand natural selection, which is the very heart and soul of evolutionary theory. He goes on to argue for "The Fact of Evolution" (the title of Chapter Two) while giving a hearing to creationist/ID people like Alvin Plantinga and Michael Behe. He follows with a discussion of the some of the problems surrounding "The Origin of Life" (Chapter Three). He ends the book with chapters ten, eleven and twelve on "Philosophy," "Literature," and "Religion." There is some interesting material on the epistemological ramifications of biological evolution and whether we can construct some sort of morality from Darwin's blueprint. (No!) He analyses Ian McEwan's contemporary novel, *Enduring Love* (1997) in terms of its relationship to Darwinism, and he dregs up some raucous anti-Darwinian quotes from George Bernard Shaw and recalls Samuel Butler's *Erewhon* (1872) amid a gloss about the differences between machines and organisms. In-between there is some interesting pro and con about familiar issues within evolutionary theory.

Ruse's style is a little long-winded and (typical of philosophers) weighted down with careful qualification—yet, at the same time, he is capable of wit and sharp expression. This is the sort of book that will appeal to those readers with philosophic, literary and historical inclinations. This is not a book for biological scientists or for those looking for cutting edge discussions of problems in evolutionary theory.

Three things:

(1) I am not enamored of the term "Darwinism" as a synonym for biological evolution. It is too restrictive both in terms of the ideas and manifestations of

evolution, but also in terms of the historical record.

(2) While Ruse understands that the facts about our biological nature revealed by evolutionary biology do not necessarily support any kind of "ought" about how we should behave, he doesn't seem to realize (see pages189-193) that we can understand and forgive on a biological level while on a societal level we must throw the violent criminals into jail. There is no "very fine line" between knowledge of our biology and our ethics, and "Darwinian approaches to humankind certainly" do not "cross that line." (p. 193) "What is" is one thing. "What should be" is another. People crossing that blunt and clearly marked line do so on their own.

(3) Little light and certainly no resolution are brought to bear on the problem of evil and free will that Ruse addresses beginning on page 284. He brings the matter up because some people think that seeing the world from the view of biological evolution somehow supports evil in the world and argues against free will. The problem of evil is the result of the belief in a personal God who is all powerful and at the same time all compassionate and has nothing to do with biological evolution. Furthermore, the question of whether we have free will or whether it is just an illusion we cannot help but believe, is a philosophical problem and not a scientific one.

Ryan, Frank *Darwin's Blind Spot: Evolution Beyond Natural Selection* (2002)
*****
Exploring the importance of symbiosis in evolution

What Frank Ryan demonstrates in this book is that evolution by symbiosis has been a "blind spot" for evolutionists since the time of Darwin, and even today is greatly underestimated by the Darwinian establishment as a force in evolutionary change, especially in speciation.

Ryan, who is an expert on viruses having penned such well-received books as *Virus X: Tracking the New Killer Plagues* and *The Forgotten Plague*, begins with some interesting history from Darwin's time showing that Darwin did not (and could not, to be fair) appreciate the role symbiosis plays in evolution. Indeed Ryan demonstrates that the process of symbiosis, and its sister processes, parasitism, mutualism and disease, itself has been misunderstood. A relationship between species may begin as parasitism (or disease) and eventually evolve into a symbiosis. This experience between species has been going on since before there were multi-cellular organisms, and is a feature of every species in existence. All species interact with some other species in symbiosis.

This central realization of the book leads to something like a new way of looking at evolution. Natural selection is still a factor, but not necessarily the major factor anymore. This is implied in the discovery not too many years ago that the mitochondria that inhabit the cells in our body are almost certainly the remnants of a once free-living bacterium that, long ago in the primeval soup or near an undersea volcanic caldron, entered a cell and stayed. We are then the product of symbiosis, which may have begun as one cell invading the other, and over the eons turned into a domestic living arrangement with the invading cell providing power to the larger cell as that cell protects and feeds the symbiont that is now earning its keep.

How eye opening this conception is! Imagine the planet filled with life forms that are composed of a dozen, or perhaps hundreds of similar arrangements made over the eons. This is evolution not by gradual steps but evolution by saltation, with a new species arising almost (geologically speaking) immediately. Such a conception would explain many of the gaps in the fossil record.

Ryan builds a strong case. Along the way he looks favorably upon James Lovelock's Gaia hypothesis (one of my favorite modern ideas) and explores the role that viruses have had in gene transfers and speciation. He contrasts the neo-Darwinian reductionists (Dawkins, et al) with a different breed of evolutionary biologist including Lynn Margulis, Erik Larsson, Luis P. Villarreal, Kwang Jeon, John Maynard Smith, Eors Szathmary, and others. He also recalls some scientists who pioneered the ideas of symbiosis but never got the credit they deserved and were virtually ignored by the Darwinian establishment. It is surprising to see how "blind" the evolutionists were and how hard it was (and is) for new ideas to gain a foothold in any scientific community. But that is the way it should be: a new idea is just a notion until it finds collaborative support by being tested scientifically.

The Gaia metaphor is perhaps the ultimate expression of symbiosis in that it involves the entire biosphere. Ryan recalls Lovelock's view that our planet with its atmosphere and self-regulating processes represents "an emergent property" of life "tightly coupled with the physics and chemistry of the Earth's environment." (p. 112) This view has yet to gain full acceptance in the scientific community, but as knowledge of the symbiotic and cooperative nature of life (instead of an emphasis on the competitive nature) becomes more widely known (and as the old scientists retire!) I think that will change. Ryan makes it abundantly clear that (to recall an expression I either dreamed up or cribbed from somewhere) "Everything works toward a symbiosis."

One of the bugaboos in natural selection has been the idea of group selection. This has been debated for many decades, but it is becoming increasingly obvious (and Ryan strongly supports this view) that group selection is a reality. Ryan reports on the work of David Sloan Wilson and Elliott Sober, who used mathematic models to demonstrate how group selection might work. (p. 255) I have argued elsewhere for group selection so I won't go any further than to note that the biosphere that survives versus the one that doesn't (either through pollution, madness, lack of foresight, inability to ward off incoming disasters, etc.) is *selected*.

The most controversial idea in this book may be Ryan's insistence that natural selection should be seen as "an editorial force" acting upon what he calls "the creativity of the Genome." (p. 265). He has German biologist Werner Schwemmler suggest a balance by noting that the "combination of the two explanations (Darwinian gradualism and symbiotic saltation)" together progress "toward a unified theory of evolution." If this is correct, the way we view biological evolution is going to change dramatically in the years to come.

Ryan makes a distinction between endosymbiosis and exosymbiosis, the former involving one genome living within another, the latter pertaining to relationships such as that between pollinating insects and plants. I want to add that the exosymbiosis between humans and our crops and domestic animals has been the essential factor in our becoming a new sort of creature, one that evolves culturally rather than biologically, and will within a twinkling of time evolve into something that we cannot yet envision because of this rapid cultural evolution. Perhaps, as some have suggested, we will form a symbiosis with our intelligent machines and let Darwinian evolution edit the result.

Bottom line: an exciting book, challenging and filled with information and ideas.

Smith, Cameron M. and **Charles Sullivan** *The Top Ten Myths about Evolution* (2007) *****
Excellent and very readable intro to evolution

"Misconceptions" would be a better word in the title than "myths," but no matter. I give this very readable book five stars because of its educational value. Reading this book is an excellent introduction to the basis and the ideas of Darwinian evolution, as well as providing talking points to refute the specious argument of creationists and "intelligent designers."

The first misconception is in the common interpretation of the phrase "the survival of the fittest." The Darwinian jungle is indeed a jungle (and a savannah, an ocean, a river, a desert, etc.), but the key to survival—being "fit" and successfully reproducing—usually has a lot less to do with how sharp your claws or how great your physical strength. Rather it has to do with how well you can make a living in the environment you find yourself in. Fitness implies such things as a good immune system, the ability to co-operate with other living things, perhaps the ability to eat a wide variety of foods, or an abundant food that will not disappear, and so on. Being able to kick butt big time is probably not a good example of fitness.

Second misconception: "It's just a theory." Yes, and a tiger is just a cat. Or, would you believe that it is very remotely possible that the earth is not round. Or, yes it is possible that only I exist and I am just dreaming up all this stuff. Actually, evolution is as much of an established fact as any theory can be. A theory, by the way, as used by scientists, isn't just an unproven idea. It is "a logical, tested, well-supported explanation for a great variety of facts." The "theory" of evolution is supported by the fossil record and the analysis of the DNA of living organisms. It is demonstrated in our lifetimes by the adaptation of microorganism such as disease bacteria. And perhaps even more importantly, its three main processes of replication, variation, and selection, remain the basis of biological understanding in a host of sciences from medicine to ecology.

Third: There is a ladder of progress (the "great chain of being") from the most primitive to the most advanced organisms (from microbes to us!). Actually the idea of progress is purely an anthropomorphic one; and the idea that evolution has some goal, ditto. Evolution is eternally a phenomenon of the here and now without any concern for the future. True, organisms have become more complex, but that is only because they couldn't have gone in the other direction! A random walk away from a wall will show, as time passes, footsteps at a greater and greater distance from the wall.

Fourth: there is a missing link that is missing. There are intermediate forms that have been discovered; and more will be discovered in the future. The fossil record is necessarily limited since very, very few of the organisms that have ever existed are fossilized. Furthermore, the transformation from one species to the other is not from one fixed type to another but from the observation of a living thing at one moment in time to the observation of another very similar living thing at another moment in time.

Fifth: Evolution is random. Mutations are random, but changes in species are anything but random. The changes are sculptured by the environment.

Sixth: People come from monkeys. We had a common ancestor with chimpanzees some six million years ago, and millions of years before that we and modern monkeys had a common ancestor. Actually if you go back far enough we are descended from pond scum. And so what if we were descended from monkeys? Some people seem to think that our close relationship with other animals is somehow demeaning. Silly.

Seventh: Nature is in perfect balance. Truth is nature is in constant flux. Balance is in the eyes of the beholder. The earth's ecological balance is an ever changing, temporary thing. At one time the "balance" was characterized by most of life finding oxygen poisonous. At another time the balance was a "snowball" earth. Who knows what the future balance will be?

Eighth: Creationism disproves evolution. Creationism is really just a kind of fairy tale, a mythology that appeals to the need of some people to feel close to their idea of God. It's a way of giving a spurious meaning to life.

Ninth: Intelligent Design is science. Actually that would be unintelligent design, and it is not science at all. Instead, ID is creationism in a tux, as some wag put it. The key misconception of ID is that we or any organism was designed. Organisms grow; they evolve. If they were designed by an intelligent designer, one imagines that they would not have as many flaws. ID is a political movement that attempts to acquire the power and prestige of science. It's a yearning for the authoritarian rule of the Dark Ages.

Ten: Evolution is immoral. Evolution is of course amoral or non-moral. What is, is from a moralistic point of view, not necessarily what ought to be. Is does not imply ought. Morality is a human idea. By the way, people who understand evolution are just as moral, or even more so, than the followers of e.g., Jerry Falwell, James Dobson and George W. Bush.

Stamos, David N. *Evolution and the Big Questions* (2008) *****
Evolutionary psychology from a philosopher's point of view

There are nine chapters dealing in turn with knowledge, consciousness, language, sex, feminism, race, ethics, religion, and the meaning of life. Professor Stamos is a philosopher as well as an expert on biological evolution as can be seen from the choices he makes on which to focus. Most evolutionary biologists would not focus on epistemology, for example, as Stamos does in the first chapter, nor would they suppose that somehow evolutionary biology could help us to find an answer to the meaning of life. But it is at the heart of Stamos's endeavor to show how evolutionary biology and the relatively new science of evolutionary psychology can at least give us a new perspective on some of these ancient questions.

The bête noir of the book is the Standard Social Science Model (SSSM) which Stamos finds not only inadequate but so riddled with politically correct and socially correct notions that it needs to be set aside. That the SSSM still holds sway in the humanities and social science departments in many of our colleges and universities is a testament to how far we haven't come. In chapter after chapter Stamos takes issue with the SSSM, applies ideas and knowledge from evolutionary biology and comes up with not only new perspectives, but in some cases enough evidence and rationale to destroy cherished conclusions. Particularly startling and revolutionary are what evo psych has taught us about religion, sex, consciousness, ethics and morality.

Some quick examples: In the case of religion, evolutionary biology has demolished the argument from design (creationists and Intelligent Designers notwithstanding). In ethics and morality we now know that morality does not come from religion but from the very nature of the beast—ourselves. The question of consciousness can be shown to be a product (or byproduct) of the evolutionary process depending on how it is understood or defined. As for sex, we now know, thanks to evolutionary psychology, that men and women have differing strategies which lead to differing behavior patterns, such as males seeking lots of reproductive tries while women try to find a male with resources. Stamos shows that the radical feminist model which claims that there are no significant sexual differences, that human behavioral differences are due to socially constructed gender roles, is mostly mistaken.

If you doubt any of this or are new to evolutionary psychology, I strongly suggest you read this exciting and surprisingly readable book. It will be an eye opener. It is clearly one of the best on the subject that I have read, and I've read dozens. Stamos is by temperament a teacher and has aimed the book at college undergraduates and a general, educated readership. His technique is to show how findings or implications from evolutionary biology relate to the big questions and then to consider objections from both experts and non-experts, contemporary and historical; and then to find a consensus or to show why there is no consensus. He is not shy about disagreeing with the heavyweights in biology or other fields nor does he mince words when taking the mistaken to task. He has the courage of his convictions and isn't afraid to criticize sacred cows. He has little patience with the kind of political correctness that leads to falsehoods. He believes in scientific truth as far as it goes and denigrates the scientific relativity of postmodernism. He is passionate and very well informed. I found the entire book fascinating, although a bit obtuse in parts. (But of course that may point to a failing on my part!)

Stamos is thorough about references, using footnotes and citing opuses, while being clear when he is giving his own opinion. His 17-page bibliography ("References") is an excellent source for further reading from which he has culled some very nice quotes as well as some telling interpretations. For example here is how he explains the official position of the US National Academy of Sciences that religion and science are compatible—a position that Stamos finds untenable: He employs William Provine, who "suspects 'intellectual dishonesty,' [is at work] just as every member of the Congress of the United

States professes to be deeply religious but surely is not. What is behind it all, says Provine, is politics motivated by the fear of a public uproar if the truth be told, votes in the case of politicians, public funding in the case of scientists." (pp. 217-218)

I think that is exactly right.

And here is Stamos on human caused pollution and destruction of habitat: "...many ecologists view the human species not as the pinnacle of evolutionary progress but as a cancerous growth or parasite, one that is destroying its host, the biosphere." (p. 217; he credits Eugene P. Odum and William H. McNeill for this metaphorical point of view.)

Stamos even goes so far as to argue against free will in humans. He reminds us of the famous experiment by Benjamin Libet which showed that when people make a conscious decision to do something, such as lift a finger, they do so half a second *after* the neural activity correlated with the willing takes place. (p. 49) We have the illusion that we are making decisions, and we live happily in our ignorance. The illusion of being in charge is part of what consciousness is all about. Stamos asks, what does consciousness do? How is it adaptive? If it gives us the illusion of being in charge, perhaps that in itself is adaptive. Someone who feels he has no say in something, who feels his efforts have no importance, who feels he is at the mercy of things he can't control may not be as effective in dealing with the environment as the man who feels that he is in charge—even if that feeling is an illusion.

Stanovich, Keith E. *The Robot's Rebellion: Finding Meaning in the Age of Darwin* (2004) *****
Partially a development from the work of Richard Dawkins

This book is largely about what psychologist Keith Stanovich sees as the disconnect in the postmodern world between "maximizing genetic fitness and maximizing the satisfaction of human desires." (p. xiii) On the one hand we have the "replicators," the genes that blindly seek only their replication. On the other hand we have the vehicle (the phenotype), i.e., "us," which carries the genes, which Stanovich believes should seek its own happiness. He sees our brain as composed of two overlapping, but sometimes divergent, systems.

One, the more primitive, he calls "The Autonomous Set of Systems" (TASS) and the other he calls an "analytic system." He calls this having "two minds in one brain."

The autonomous system is held on a "short leash" by the genes while the analytic system is on a longer leash; that is, TASS reacts to events in the environment almost automatically in close concert with the dictates of the replicators while the analytic system is more removed from innate drives and can analyze situations rationally and can act in terms of what is good for the vehicle rather than what promotes the replication of the genes. Note that these systems usually are in agreement and react to the environment in the same way. Threats to the well-being of the vehicle from predators and other dangers signal the same avoidance behavior. However, sometimes there is a conflict. The example that Stanovich uses is TASS's need to flirt with the boss's wife, which might increase the replication of the genes, while the analytic system realizes that such behavior probably goes against the best interests of the vehicle (possible loss of job, etc.). Following the counsel of the rational analytic system instead of the urgings of TASS is what Stanovich calls "maximizing goal satisfaction at the level of the whole organism." (p. 64)

The title of the book comes from Richard Dawkins (and indeed this book is written in partial reaction to and in concert with Dawkins's ideas) who called organisms "survival machines" and "gigantic lumbering robots" in his famous opus, *The Selfish Gene* (1976). Stanovich wants to free us from the dictates of those selfish genes and so has constructed a "robot's rebellion." He believes we can use our rationality (our analytic system) to override the sometimes self-destructive inclinations of the more primitive set of brain systems. Stanovich is preeminently a rationalist and believes that right thought leading to right behavior will lead to a more fulfilling and happier life for the "robots." We need to be on the long leash from the genes, not the short leash, is his idea.

A strong point that Stanovich makes very well is that in the information societies of the modern world many of the talents that served us well in the Environment of Evolutionary Adaptation in the Pleistocene are "worthless" when (e.g.) trying to use "an international ATM machine with which you are unfamiliar" or when "arguing with your HMO about a disallowed medical procedure." (p. 124) He argues strongly that corporations and governments, through their advertising and propaganda, have become very good at exploiting blind spots in our more primitive brain systems and getting us to do what is good for them and not necessarily good for us. I think this is correct, and that those of us who can see how the players in the modern economy are trying to use us for their benefit will avoid most of the more obvious traps and thereby increase our standard of living and presumably our chances for happiness.

Stanovich devotes a chapter to criticizing evolutionary psychologists for failing "to develop the most important implication of potential mismatches between the cognitive requirements of the EEA and those of the modern world," as he carefully phrases it on page 131. Nonetheless the psychology presented here is mainly a synthesis of cognitive psychology, brain science and evolutionary psychology and as such represents the latest in our attempt to understand ourselves.

He also devotes a chapter to the effects that another kind of replicator, the meme, has on our lives. I don't have the space to go into his ideas about memes and their implications, but I want to say that from my point of view the word "meme" is an approximate neologism for the word "idea." However, I think that it is a useful coinage and, like Stanovich's mind dualism, facilitates a new way of looking at and talking about how our brains work.

While I think this is an extremely interesting book that goes a long way toward showing us the sort of thinking that characterizes postmodern psychology, I must point out that Stanovich's mind dualism is a construct that, while based on his interpretation of recent findings, is nonetheless just that: a construct that will be refined as time goes by and eventually overturned for a new construct. As always in science we are increasing our understanding and expanding our knowledge as we move toward a final understanding that will most likely always lie tantalizingly in the distance.

Zahavi, Amotz and Avishag *The Handicap Principle: A Missing Piece of Darwin's Puzzle* (1997) *****
Best book on evolution in many years

Why does the peacock grow that tail? Why does the springbok leap straight up into the air when it sees a predator? Why do people behave heroically? The handicap principle answers these questions, eloquently, simply and with an overwhelming sense of conviction. The peacock is advertising his fitness. He is saying to the female in essence, I am so fit I can carry around this cumbersome adornment and still scratch out a very fine living. The springbok is saying to the predator: don't even think about going after me. I am in such good shape I can waste energy jumping up and down and still have plenty of reserves to outrun you. Save us both the bother and go after someone weaker. (By the way, the springbok jumps straight up instead of sideways because by jumping straight up its performance can be effectively judged by a predator from any direction.) And the man who dives into the swiftly flowing river to save a drowning child is actually advertising his fitness and improving his station in society. He is so fit he can take chances that others dare not. He's the man the women want to mate with.

The Handicap Principle thus is about signals, signals between prey and predator, between one sex and the other, and between the individual and the group. The purpose of these signals is to display in an unequivocal way the fitness of the signaler. Note that such signals have to be "fake proof." They have to be what the authors call "reliable." An animal that can't run fast and has limited resources of energy can't waste them jumping in the air. It needs to get going immediately or to stay hidden if it is to have any chance of survival. A man leads with his chin. That's a signal that he's confident. When men had beards it was a little dangerous to stick your chin out since the other guy might grab your beard and you could be in trouble. People demonstrate wealth by wasting money. This is a "reliable" (if ugly) signal because without an ample supply of money, you can't afford to waste it.

Part of the beauty of this book comes from the personality of the authors, who spent a large part of their lives studying little babbler birds in Israel. I feel I know these little social birds just from the loving descriptions in the text. One can see that even though the Zahavis made their discovery of the handicap principle in 1975 and waited almost two decades before it was generally accepted in the scientific community, they harbor no bitterness, nor is their tone at all gloating. They come across as hard-working field scientists who love their work and nature.

Besides being full of exciting and original ideas, *The Handicap Principle* is also extremely well written. Each sentence is clear and to the point without the burden of unnecessary jargon or the wordy clumsiness sometimes found in such books. Amotz and Avishag Zahavi took great pride in effectively communicating their ideas to a wide audience. Additionally there are scores of exquisite, loving little black and white drawings by illustrator Amir Balaban of animals, birds, insects and people, etc., illuminating the text.

If you're interested in evolutionary theory, this is a book not to be missed. As Jared Diamond says on the cover, "Read this fine book, and discover what the excitement is all about."

Zimmer, Carl *Parasite Rex: Inside the Bizarre World of Nature's Most Dangerous Creatures* (2000) *****
Ugh, yuck and yike!

The prevalence of parasites in the earth's ecosystems is perhaps the worst of the dirty little secrets of life. According to science journalist Carl Zimmer, the majority of animal species are parasitic (p. 187). Bambi and Mickey Mouse had 'em, the sparrows outside your window are riddled with them, and you and I have them. What you'll read about here is how creatures insinuate themselves into other creatures and take control of them; how, for example, a wasp inserts its eggs into a caterpillar and how the larva eat the caterpillar's flesh, and other equally ghastly processes. The result is a view of life without rose colored glasses or any cute Disney characters.

I had a dream one night before going to the dentist. I dreamed that there were scores of little white noodles like spaghetti coming out of my gums and tongue when I pressed them. Initially I thought this was a dream of the grooming instinct. But after reading Carl Zimmer's creepy but utterly fascinating book, I believe this primeval dream may have been a dream of parasites.

In a sense this is the scariest book I have ever read. I had to put it aside twice because the horror depicted on its pages was affecting my usually buoyant state of mind. Make no mistake about it, however objectively we may try to view our fellow creatures, it is impossible (at least for me) to see parasites as anything other than ugly and despicable. (I'm working on it, however.)

But Zimmer has an important purpose in writing this book aside from scaring us. He makes it clear that we cannot understand how an ecology works without understanding the role parasites play in that ecology. For example on page 111 we have wolves choosing to attack a moose that is slow and wheezing, a moose riddled with tapeworms, tapeworms seeking their final host, the wolf! The wolves are led to choose the infected moose perhaps by a scent in the moose's breath, created by the tapeworms. "The thinning of the herd is an illusion, not the service of the predator but the side effect of a tapeworm traveling through its life."

In Chapter 3, "The Thirty Years' War," Zimmer shows how the immune system fights against parasites. It is an excellent exposition on how the immune system works, and one of the highlights of the book. In the chapter "Evolution from Within," we see parasites as a driving force in evolution. The idea that sexuality began as a way to fight disease, the so-called "Red Queen" hypothesis is presented. Zimmer shows how being different instead of a clone of the mother (asexual reproduction) can lead to characteristics that foil parasites. There are sixteen pages of glossy photos of parasites, several showing the grotesque heads of tapeworms in intimate detail. There are photos of a crustacean parasite that invades a fish's mouth, eats its tongue and then takes the place of the tongue.

A question that might be asked is, what is a parasite? Certainly in biology a parasite is different than a symbiont, which is distinguished from a predator, etc. Zimmer gives Richard Dawkins's definition that "Parasitism is any arrangement in which one set of DNA is replicated with the help of—and at the expense of—another set of DNA" (p. 126). By this definition perhaps humans are parasites on Planet Gaia. Zimmer suggests as much on page 245 adding that "There's no shame in being a parasite...But we are clumsy in the parasitic way of life." He explains that expert parasites do no more harm than is necessary. "If Gaia had an immune system, it might be disease and famine" to keep "an exploding species from taking over the world. But we have dodged these safeguards with medicines and toilets and other safeguards..."

Zimmer ends the book on a rather cheery note: parasites as the canary in the coal mine. He argues that parasites are not only an indication of ecological health (plenty of parasites suggests a healthy ecosystem; a drop in their numbers suggests trouble, perhaps from pollution), but are vital to the ecology by keeping animal populations in check (pp. 241-243). He also gives some idea of how parasites might benefit us more directly, such as a fungus that invades insects as "the source of cyclosporin, an important antibiotic" or the use of blood-clotting molecules produced by hookworms as blood thinners in surgery (p. 238). Although there have been failures in the use of parasites to control insect populations, and Zimmer recounts two or three, there is also the story of how a parasitic wasp imported from South America saved the African cassava crop from mealybugs (pp. 220-228).

As others have noted, reading this book will forever change the way you view the natural world, and might make you cancel that trip to the tropics.

Chapter Six

Evolution and the Future

Some of these titles might more properly be termed futurist works rather than works about future evolution since they are mainly about cultural evolution as opposed to biological evolution. However from my point of view cultural evolution is a subset of biological evolution which itself is a subset of cosmic evolution.

In any case the works reviewed here are without exception fascinating.

Baldi, Pierre. *The Shattered Self: The End of Natural Evolution* (2001) *****
Speculative science of the highest order

"We do not know who we are, but we know enough to know we are not who we think we are."

This quote on page xiii (from Baldi himself, I imagine) sets the tone for this extraordinary book which is an excursion into the future of ourselves and the world we are making. The emphasis is on biology, genetics, computer science and information technology including brain science and how discoveries in these fields are changing our lives and our very concept of self.

What Baldi does so very well is to take current tendencies in these fields to their logical conclusion, and to look fearlessly at the results. This will be unpleasant reading for some, especially for those who see our species as fixed, a permanent endowment from a supernatural being. Baldi's general point is that we are forever changing. With the end of natural evolution upon us, so that we feel the full force of cultural evolution, the pace of change is rapidly accelerating. The result of this will be that, come some not too distant future, we will indeed be something strikingly different from what we are now.

In progressive religious circles it is said that we are "becoming." Usually it is not said *what* we are becoming, because that is not known. Baldi makes the same statement from the viewpoint of science. Through the rapidly accelerating power of culture evolution, we are in the process of "becoming" that which we cannot as yet clearly see.

"[O]ur notions of self, life and death, intelligence, and sexuality are very primitive and on the verge of being profoundly altered...It is this shattering...that forms the central thread of this book." (p. 3) The idea that "each of us is a unique individual delimited by precise boundaries" is wrong. With the advent of technologies stemming from science we will see that these boundaries are artificial as we are cloned and joined with silicon parts and machine intelligence, and as we become more and more attached to the Internet.

This is Aldous Huxley's *Brave New World* for real, but Baldi's is not a dystopian vision, rather his is a vision of great hope and excitement. He believes that "genomes, computations, and minds are...continuous entities, both in space and in time," and that we, individually, "are just samples of this continuum." (p. 4)

This is a startling view of the world and ourselves, an exciting one that promises gargantuan changes to come. Whereas many people are dismayed at the prospect of cloning human beings (and our government is currently against it), Baldi celebrates the prospect, writing that cloning, "At a minimum...provides a form of genetic immortality" that "may be reassuring" to some of us. (p. 61) He asks, what is "the chance of having twin Einsteins by natural means...?" He answers that it is a 100-billion to one shot, but "trivial to replicate with cloning techniques." Baldi even dares to mention the prospect of cloning for spare parts. (This might be called "partial cloning.")

Baldi believes that someday human intelligence "might be viewed as a historically interesting, albeit peripheral, special case of machine intelligence." (p. 113) He sees our neurons in direct interface with silicon intelligence, our memories and computational powers greatly enhanced.

He is trying "to imagine what can be done with intelligence and other faculties several orders of magnitude beyond our own." He writes: "Almost by definition, what can be done is in a blind spot that our brains cannot really see." (p. 114)

It occurred to me while reading this that by gradually replacing our brain cells...etc., we would never "die," at least not knowingly. We would change, perhaps drastically, over a period of time, but our subjective experience would note only small changes similar to the experience of watching grass grow. This would be a sort of death-defying immortality (with evolutionary change), the sort of thing that might work with human beings. Note how this fits in with Baldi's idea that we are not who we think we are. Indeed, we cannot really know who we are in a definitive sense. We can only know that we are part of a larger process.

I might note that in this sense science and religion are in the process of merging. Religion is the phenomenon of belief; science is the process of "knowing." At some point, off in the distance, there is a whole universe that invites belief without our having an ability to know. This is where science and religion merge.

Boulter, Michael *Extinction: Evolution and the End of Man* (2002) ****
Dense, somewhat difficult, fascinating

While I think that paleobiologist Michael Boulter is certainly correct in his assertion that we are going to go extinct, as all creatures eventually do, I don't think we will go the way of the mammoth or the giant sloth or the Neanderthal. Our exit may very well be totally unique. We may go the way of the dinosaur, of course, our world obliterated by a cosmic catastrophe, or we may blow ourselves up, and then watch the survivors die out in the ruins. But more likely we will pass away quietly as our culture transforms us from what we are now to creatures that are partly the result of genetic engineering and partly the result of mechanical ingenuity, until one day we may notice that we are so different from the humans of the past as to be an entirely different species.

But Boulter is not concerned here with cultural evolution. He is looking at the biological evolution of life on earth primarily through the fossil record and in particular through Fossil Record 2, a huge database that he has studied extensively. His theme, despite the book's title, is the diversity of life, the radiation of living groups, etc., and how an understanding of that diversity through an analysis of the fossil record can shed light on the evolutionary process. He analyzes the growth of life's diversity after the major catastrophic events in the earth's history and plots curves and comes to the conclusion that biodiversity is an example of exponential growth, and that the phenomenon of evolution is another example of a self-organized system (such as sand piles and the weather) driven by "power laws and pink noise." (p. 125)

Some of the interesting conclusions that Boulter comes to along the way to forecasting our extinction is that we probably did do in the Neanderthal. (He lists "selfishness" as one of our distinctive traits that the Neanderthal apparently didn't have enough of.) And yes, we wiped out the major fauna of North America within a thousand years or so of our arrival from across the Bering Strait. In fact, we are now living through a period of mass-extinctions, in particular of large mammals, and we are a major factor in those extinctions.

My problem with this book is that it is sometimes hard to follow Boulter's argument since he is not as direct as he might be. Then again it may be that I need to read more carefully! At any rate, the fact that biodiversity follows an exponential curve until it hits a catastrophic event is certainly one of his points. And that evolution is an example of a self-organizing system like that of a sand pile, and behaves in similar ways with large changes occurring less often than small changes, etc., is another. Do "groups of animal and plant Families follow clear rules in their origin, expansion, peak diversification and eventual extinction?" is a question he asks. (p. 124) His answer is yes, and the pattern can be traced. He adds that "extinctions are an essential stimulus to the evolutionary process." (p. 183)

The "new idea" (as he terms it, p. 182) that mass-extinctions come from "within" as a feature of self-organization does not seem convincing to me, although it is certainly intriguing and worthy of further study. He writes: "So modern man is kicking the sand pile and causing a severe avalanche that only started to crash down at the end of the last ice age...the fundamental cause continues: human aggression. The first phase was our killing other mammal species...then through human history our killing of one another."

But is it only a temporary irony that today there are more humans on this planet than ever before?

Aggressive we are. And we kill each other with an amazing abandonment, but have such actions led us toward extinction? The evidence is all to the contrary mainly because our reproductive abilities and our ability to exploit planetary resources outweigh our murderous tendencies. And besides the cause of at least some of the great mass extinctions of the past (huge meteorites) clearly came from without.

Boulter sees small animals inheriting the earth after we are gone. He notes (p. 193) that "insects and birds are still at the early stage of high diversification." What this means is that a group of animals that is continuing to diversify (continuing to grow in the number of species) will be safe from extinction until the diversification slows. This is a nice scientific understanding, but what it says to me is that a successful body and behavioral style (e.g., a Family or order or some other classification of organisms) is less likely to go extinct than a less successful one. One might say, QED.

He speculates (his terminology, page 176) that "our system is in free fall, out of control." We won't need "nuclear weapons," he posits, "or the inventions of science fiction writers." We are "doing very well...just with our use of fossil fuels." Exactly what he has in mind here is not entirely clear. Does he mean that we will pollute ourselves to death?

Elsewhere he writes about global warming, caused in part by our burning of fossil fuels, but advises that fluctuations in temperature are common, and that for much of the history of life on this planet it was hotter than it is now, and that, in fact, for 250 million years from before the P-Tr mass-extinctions until the Miocene there was no frost on earth. (p. 113) Furthermore, "between AD 900 and 1300 cattle were farmed in Greenland and the French tried to embargo English wine." (p. 122)

In short, this is not a text for the casual reader. It is dense, and in places, technical. But what Boulter has to say is worth the effort.

Garreau, Joel *Radical Evolution: The Promise and Peril of Enhancing Our Minds, Our Bodies—and What It Means to Be Human* (2005) *****
Dense exploration of the technological explosion to come

This is about the so-called GRIN technologies: Genetic, Robotic, Information, and Nano. Properly speaking the title should be "Extreme Cultural Evolution," or perhaps "Accelerated Technological Evolution." "Radical" is used here in the sense of "extreme." Regardless of what we call it, for better or for worse, we will be enhancing our minds and bodies and changing the life forms around us, especially those we use for food. In fact we have already done so through computers, surgery, artificial limbs, genetically engineer agricultural products, etc. The difference to come is all about the acceleration of change coming from these technologies.

What happens when your daughter's brave new genetic endowment gives her a prodigious memory and makes her smarter, prettier, and stronger than you? No problem. We love our children. Ah, but what happens when she realizes that at age eighteen she is like an Australopithecus creature compared to the new genetic and nanotechnological enhancements bestowed upon her classmates just a few years younger?

What happens is the end of the world as we know it, and most critically the end of human beings as we know ourselves. The question is, is this is a good thing or a bad thing?

Joel Garreau has several answers in terms of scenarios of the future. There is the "Heaven Scenario," the "Hell Scenario," the "Prevail Scenario," and the "Transcend" possibility. Garreau interviewed a number of experts in many fields in an effort to find out not only what the prospects are, but to count noses, so to speak, and see who's optimistic and who isn't.

Put Ray Kurzweil, author of *The Age of Spiritual Machines* (1999)—see my review on Amazon—in the camp of those who see marvelous things happening, in fact a glorious singularity of advancement. Put Bill Joy, co-founder of Sun Microsystems, in the camp of those who believe we are headed for a right awful hell on earth. And put polymath Jaron Lanier in the camp of those who think we can prevail over our creations. And put Michael Goldblatt of the US military's Defense Advanced Research Projects Agency (DARPA) in the platoon of happy warriors just having fun with the prospect of new and more amazingly advanced weaponry (or defenses from weaponry).

After reading this dense and fascinating book I have a few observations. First, regardless of whether we like it or not, or whether Luddites and social conservatives manage to slow down or even halt some of the research, nothing but nothing is going to stem the tide, or alter The Curve, as Garreau calls the shape of things to come. If we don't do stem cell research or explore replicating nanobots, you can be sure that somebody else—in Korea, in China, in Russia, even in Pakistan—will. Any nation or culture that chooses to not explore these brave new worlds will be in danger of not only being left behind economically and militarily, but in grave danger of living a sub existence like that of pets or zoo animals.

There is some debate about this point. Garreau explores the idea that nothing will stop the tsunami and does find some people who think we can put up a wall or at least quiet the rampaging waters. Still others are asking, why should we? Think-tanker Francis Fukuyama, author of *Our Posthuman Future: Consequences of the Biotechnology Revolution* (2002)—see my review at Amazon—believes there is something precious in humans as presently constituted. He is fearful that we will lose that human nature through biological engineering. Personally, glancing at the history of human kind, I think that human nature could use some altering, and indeed believe that unless human nature does change, we won't be around much longer. Fukuyama believes that, were we to become as immortal as the gods, we would stagnate. He "doesn't think immortals will ever have a new idea again" and only the death of people allows new ideas to take root. (p. 163)

What if we do conquer all and end up with this so-called heaven on earth? What will it consist of? Will we pursue endless delights from brain chemistry? Are we creatures ruled by the gods of pleasure and pain, or is there some transcendental aspect to us? Garreau explores this question near the end of the book with help from Martin E.P. Seligman's three levels of happiness: "the pleasant life, the good life, and the meaningful life." Here I think Garreau, along with Seligman is whistling Dixie in the dark. The "meaningful life" is what? According to what I could gather on pages 261-262, the "meaning consists in attachment to something bigger than you are." Seligman finds such attachment in various activities from raising children to saving the whales to being a terrorist. I think a more lasting attachment may be to something like exploring the cosmos.

But would humans really have sufficient desire to do that? Recalling some famous dystopias from literature, H.G. Wells's *The Time Machine* or Aldous Huxley's *Brave New World*, for example, I suspect that creatures such as ourselves (as currently constituted) can only exist in environments not that far removed from the savannah. Cities are tough enough for the couch potato obese of the Western world. If we gain everything our biology desires, we may become (further) degenerate and fall victim to something untoward and unpredictable. Or we may just end up examining our navels as the perfect mixture of chemicals courses through our bodies. If we conquer all and have no challenges left, what will we do? What does a perfectly satisfied and perfectly serene creature do? We don't know. Transcend human nature perhaps?

Naam, Ramez *More than Human: Embracing the Promise of Biological Enhancement* (2005) *****
Brave new world or genetically-enhanced pipe dream?

The basic thesis of Ramez Naam's book is that with our ability to shape (especially to enhance) our biological nature through the tools of our culture—in particular, genetic engineering—we will transform humanity into "a plethora of forms," which will eventually result in thousands if not millions of new species. Naam contends that we will spawn "a new explosion of life as sudden and momentous as that of the Cambrian explosion" some 570 million years ago. (p. 233)

That's the upside. What is also possible (although Naam does not dwell on this) is that with biological enhancement tools that are presently coming into discovery and use, we may transform ourselves into beings who will have satisfied their every desire, and with that satiation, have put an end to desire. The result may very well be the end of human evolution, biological or cultural. And following that, the end of the species that began as a big-brained walking ape six million years ago.

Or none of the above.

This is the exciting part. We have no idea where cultural evolution is going to take us. We have no idea whether we will develop the ability to stave off natural disasters (rogue comets; nearby supernovae; unstoppable pathogens) or overcome our propensity to self-destruction in the form of perpetual war or the poisoning of our environment. Yet, modern Luddites and social conservatives notwithstanding, we will indeed use the tools we develop to initially prevent and cure ailments and deficiencies, and ultimately to enhance our ability to enjoy and to get the most out of life.

This is what this book is all about. Naam begins with the fuzzy distinction between using genetic engineering to heal or to enhance, and makes two telling points: (1) it is often impossible to distinguish between a procedure done as part of the healing arts, or one done to enhance our abilities; and (2) whether we like it or not, given human nature (as it now exists!) if the enhancement tools are there, promising greater intelligence or greater beauty or longer life, then we humans will inevitably use such tools. If the Bush administration or some other Luddite-mentality government tries to suppress these tools, people will just go elsewhere. And those societies that fall behind will fall very far behind. The genetically enhanced will inherit the earth, and indeed it isn't much of a stretch to imagine a future in which those who have enhanced themselves are so far in advance of those who have not as to constitute superior beings. Will the Luddites become pets?

More immediately—keeping these ideas in mind—will it only be the rich who will benefit? Naam argues—and I think convincingly—that yes, at first only the rich will use the tools to better themselves and their children, but then lagging ten or twenty years behind will come the total mass of humanity. Naam compares this process to that in the present day pharmaceutical environment in which initially the new drugs are very expensive, but after they go generic they become affordable to the masses.

There is so much in the book that I will not be able to get to even a fraction of it. So let me say that Naam has anticipated a lot of the criticism that will be leveled at his position and he has done a good job of answering it. The idea that we can somehow stop genetic engineering to save our human nature is shown as bogus since human nature is an ever evolving, ever changing abstraction. Even the concrete species itself (which is us) has changed mightily over the eons from Australopithecus to homo sapiens. And whether we lift a finger or not, we will eventually change again or go extinct. That is the main point. We cannot stop change. We cannot hope to preserve the present human "endowment." We can only hope to engage change, and with our intelligence make life better for ourselves and those to come, people who will be different from us, and going far enough into the future, very different from us.

For the here and now, Naam sees biotech and neurotech enhancements as "investments in valuable human capital." (p. 76) I believe this is the primary reason the United States must overcome the backward mentality of the Bush administration and support not only more stem cell research, but encourage a greater investment in all forms of biological engineering. If we don't we will fall behind those who do.

For others who see the ghost of eugenics in his position, Naam has an effective answer. He writes, "the only people advocating state control over the genetic makeup of the population are those who would like to see genetic enhancement techniques prohibited. The advocates of human enhancement, on the other hand, are arguing for individual and family choice, the opposite of state control." In other words, "...the prohibitionists are the ones upholding the eugenic side of this debate." (p. 166)

Naam gets very specific about the enhancements possible or at least conceivable, including brain-computer interfaces, brain implants, human cloning, electrical stimulation of the brain, pre-implantation genetic diagnosis (which takes in vitro fertilization one step further), etc. Near the end of the book, he sees us communicating not only ideas and words, but thoughts, feelings and emotions to others directly from our brains as one would communicate through a wireless network. Eventually we will have "the flexibility to do what we like with the contents of our thoughts, feelings, and imaginations..."

Since all of this may sound scary (yet exhilarating), Naam adds, "and society will respond with new social norms to guide our choices." (p. 219)

Oh, brave new world that has such things in it!

The book is fascinating. Naam has not only done his homework, he has thought out the consequences of what he has found and provided the reader with some guidance.

Palumbi, Stephen R. *The Evolution Explosion: How Humans Cause Rapid Evolutionary Change* (2001) ****
Colorful take on how we cause unwanted evolutionary change

It has become clear over the last few decades that evolution can take place much more rapidly than Darwin ever imagined. The evolution of the AIDS virus is a particularly compelling case in point, and one of the focal points of this engaging book about how our efforts to control our world can bring about unwanted evolutionary change over time periods measured not in millennia, but in weeks and months. Mostly it is microbial evolution that Harvard Professor of Biology Stephen Palumbi writes about, the AIDS virus, the bacteria that cause tuberculosis, staph and other infections, but also insects and plants, particularly the insects that eat crops and the plants we call weeds, and even fish. At the center of change is the "evolutionary engine" that is continually at work adjusting organisms to their environments. Change the environment of a creature and the creature changes to keep its fit, a never-ending phenomenon that frustrates our efforts to eradicate harmful pests and deadly diseases.

Palumbi shows how it is not enough to spray our fields of amber grain with pesticides because the pests will inevitably evolve to flourish in the new pesticide-filled environment. It is not enough to throw antibiotics at the bacteria that invade our bodies because they too will evolve to flourish. Our efforts to combat the scourges of field and body are now seen as just one half of the prey/predator, parasite/host phenomenon of co-evolution. As Palumbi phrases it, "The disease dance continues, turning to the evolutionary tune, and both players must step smartly." (p. 90) We must take the power of life forms to evolve rapidly into account and realize that they will react to our efforts. This is the evolutionary arms race, the "Red Queen" hypothesis, that keeps us (if we "step smartly" enough) and our enemies in the same place even though we are both running at full speed. This may be seen as a kind of cosmic joke at those who would find "progress" in evolution.

En route on bringing us up to speed on rapid evolutionary change, Palumbi sets some sort of record for the use of colorful language. There is some distraction as metaphors and analogies fly about like confetti at a wedding, but he is so clever that we forgive him. Some examples:

p 16: "...as unknown as the dreams of a sleeping infant."

p. 56: a trait (a recessive gene) is said to lie "dormant like thoughts on a Saturday morning."

p. 102: a virus is compared to a credit card.

p. 107: a typical viral attack on the immune system "has more plot twists than a soap opera."

p. 137: expressing the too-optimistic hopes of a five-year malaria eradication program: "...by then, surely malaria would be gone like the world's last car payment."

p. 240: "bad ideas" are compared to "anchovy daiquiris" that "live on only in a few people with fishy breath."

In short, this book colorfully illuminates one of the most significant conundrums of our time: despite our best pesticides, our most powerful antibiotics, our most clever and hopeful chemical cocktails, we are not winning the war against pests and disease. We are at best holding our own. The message of this book is perhaps we can do more if we take into account the power of the evolutionary engine, and finds ways to use it to our advantage.

Ridley, Matt *The Future of Disease* (1997) ****
Informative, balanced

In this modest little book, Matt Ridley, the eloquent and incisive author of *The Red Queen: Sex and the Evolution of Human Nature* (1993)—see my review on page XXX—not only predicts the future of human disease (it's not as bad as some would have us believe, but it's still scary) while focusing some light on the nature of pathogens and how they propagate. In the usual lively and insightful Ridley style, we learn how modern humans are more vulnerable than ever to aerosol diseases (spread through breathing, e.g., colds and flu) and sexually transmitted diseases, but less vulnerable to water spread diseases (dysentery, cholera, etc.) because of improved sanitation, or to vector spread diseases (e.g., yellow fever, plague) because of vector control, drained swamps, and fewer people living under rat-invested thatched roofs. He also explores our vulnerability to newly mutated microbes and the propagation of disease in hospitals and other niches in the modern environment. He speculates on the next great plague and where it will come from and its nature. He talks about the AIDS epidemic and the Ebola scare and compares them to past scourges. He even mentions prion contagion.

Ridley is neither overly optimistic nor needlessly pessimistic. He warns on page four that "there is no end to the struggle with disease. Infection is never going to be entirely defeated." He adds that the human population is "too gigantic an ecological niche to be left vacant." On the up side he writes that "we are on the threshold of a new age of technology" that includes the promise of DNA vaccines and molecular-designed drugs to help us fight the parasites.

Rose, Steven *The Future of the Brain: The Promise and Perils of Tomorrow's Neuroscience* (2005) ****
How neuroscience will and will not change our lives

The Future of the Brain is about how neurotechnology derived from neuroscience will attempt to change our brains, about what we can and cannot expect from science, and what we should fear. Rose is a brain scientist whose specialty is in the neuroscience of memory.

He is also a prolific writer on evolutionary biology. He is a proactive opponent of a strictly reductionist stance in biology and a stern critic of what he sees as a genocentric approach to psychology and what it means to be human. Some of his books (most notably *Not in Our Genes* (1984) written with Richard Lewontin and Leo Kamini, and *Alas, Poor Darwin: Arguments Against Evolutionary Psychology* (2000) which he edited with Hilary Rose) are more about the politics of evolutionary biology than about the science; but here Rose keeps his political views mostly in the background. The result is an informative book that helps us to understand what science is learning about how the brain works and about how it can be affected by outside agents.

After an introductory chapter he begins with the nitty-gritty of how the brain came to be and how it might be understood—from proto-cells in the pre-biotic soup to axons, dendrites, synapses and brain "structures." His theme throughout is that the brain cannot be understood except as a process continually in motion. He argues that how our brains developed cannot be appreciated through an isolated study of the genetic blueprint. Instead we must look to the brain's developmental history in interaction with the environment to determine what it is and how it works and why.

The middle chapters move from the brain to the mind, from the nuts and bolts of neurology to the experiential human being living in an environment in part created by itself. Rose touches on the "mystery" of consciousness and the paradox of free will. He finishes with some conjectures about what kinds of pharmaceutical agents are to come, what kinds of invasive procedures might be employed in attempts to combat various diseases or to cope with the effects of ageing or to help make us "better than normal." The final chapter is on "Ethics in a Neurocentric World."

Although Rose does not spell out how the mind differs from the brain—I take it he presumes a dictionary definition—much of the book is concerned with the distinction. The brain is the flesh and blood; the mind is the experience, is how I read him. I want to add that the distinction between brain and mind can be seen as similar to the distinction between sex and gender. Sex is biology. Gender is the cultural expression of that biology.

He objects to viewing the brain as composed of "modules" directed by genetic imperatives. He writes that "...life is not a static 'thing' but a process" (p. 62) We are forever changing. The Steven Rose of 30 is not the same as the Steven Rose of today. He is a different person because of what has happened to him during the ensuing decades, and how he has reacted to what has happened, and what he has learned. And if Steven Rose were somehow cloned, that Steven Rose would be different still because of the different environments—prenatal and afterward—in which he would grow.

He speaks of "patterns of activity" in the working brain. He doesn't like the use of "modules" such as a supposed "reading module" or "reading instinct." (p. 134) However it is really impossible to write about something as foreign to our everyday experience as the workings of the brain without resorting to metaphor and analogy. Something is like something else. Something is compared to something else. This is how we learn. So instead of modules, Rose employs variously, "a collection of mini-organs" (p. 149); "brain regions" (p. 157); "brain...structures" (p. 133), etc. In fact he uses the term "modules" himself on, for example, pages 149, 156, 158. Furthermore his railing against the use of our experience in the "Environment of Evolutionary Adaptation" during the Pleistocene by evolutionary psychologists is partially contradicted by his acknowledgment that we are indeed shaped by our environment as we in turn shape it. It is clear to me that where Rose and the evolutionary psychologists differ is in their perception of how much the environments since the Pleistocene have changed us. Steven Pinker, Edward O. Wilson and others think "not all that much," while Rose thinks "a whole lot." The truth, one can imagine, lies somewhere in between.

It should be noted that one of the unsolved problems in evolution is knowing how fast evolutionary change can take place. Stephen Jay Gould spoke of rapid change after long periods of stasis while others have disagreed; but no one can say how much we have changed biologically since the Pleistocene. It is known that large populations are strongly resistant to evolutionary change because mutations quickly get swamped in the huge genetic pool. My feeling is that in populations as large as ours, little evolutionary change is taking place. The environment is constantly changing, but the selective pressure usually brought about through starvation, disease, and competition from other species is really not much in evidence. And so I tend to side with those who believe we haven't changed all that much.

Steven Rose is a wise and caring man who sometimes forgets his manners when speaking about those with whom he has sharp disagreements. But in this book he is at his best and most well-behaved. Let me finish with perhaps the wisest of his observations. He is speaking of the increased "powers of surveillance and coercion available to an authoritarian state." He warns, "The neurotechnologies [now available and to come] will add to these powers, but the real issue is probably not so much how to curb the technologies, but how to control the state." (p. 302)

Ward, Peter, images by Alexis Rockman. *Future Evolution* (2001) ****
Striking images and a sprightly text

This is almost as much of an art book as it is a book on evolution. The images, photos of about 30 paintings by Alexis Rockman, mostly oil and acrylic on wood or watercolor and ink on paper, are stunning depictions of creatures, past, present and to come: an arsinotherium (a rhino-like animal), a thylacine (a doglike marsupial, extinct in 1936), huge dandelions with thick roots several feet long, rabbits and rats on hind legs like kangaroo, crows like vultures, snakes with wings, etc. The text by geologist Peter Ward is sprightly, informed, very readable, and at times even moving, as when Ward recalls his return to New Caledonia after twenty-five years.

Ward's vision, however, is not pretty. He is not looking at planet earth after humans have gone extinct as some other books on future evolution have done. He sees us as surviving for another 500 million years so that the fauna and flora that do evolve will do so with humans as probably the most significant part of their environment. Consequently there will not be any large mammals, and the most numerous creatures will be small and "weedy." They will be mostly nocturnal animals that have learned to tolerate humans, rats and insects and "escapes" from our farms and genetic engineering labs.

Ward is very good at producing striking word portraits. One is the "brown mountain" he observed flying into Mexico City (the polluted air rising above the city), and another is his fanciful creatures of the future, the "Zeppelin-iods," who have learned how to create hydrogen-filled air sacks so they can float in the air. In a particularly dystopian vision on pages 135-137, Ward's time traveler visits a garbage dump 10-million years in the future crawling with "cockroach-sized insects...[and] mammals, a few as large as cats but most rat-, mouse-, or even shrew-sized." These creatures have evolved adaptations for exploiting the garbage dump: "some with long tapered heads, others with thin ribbonlike tongues, others with blunt heads and large knoblike teeth, still other with huge batlike eyes." A pig-like creature with rats "like hairy lampreys with greedy sucking mouths" hanging from its sides appears. Overhead large crows "with brilliant plumage" dive bomb the traveler with knifelike barbs on their feet, driving him bleeding toward a tree where a hungry flock of these clever and hungry crows await. Ward also sees a great increase in the number of snakes, some with unusual adaptations to feed on the garbage eaters.

This "dyspeptic" vision, like some of the other visions in the book, is calculated to shock and revolt the reader, but just how likely is it to come to pass? On the one hand it would seem, not very, since we are already recycling away from garbage dumps in many places in the world. On the other hand, if we consider that we, as domesticated creatures ourselves, may be getting dumber, this scenario might seem more likely. (See page 105 where Ward references neu-rologist Terry Deacon as noting that "all domesticated animals appear to have undergone a loss of intelligence compared with their wild ancestors.") My feeling, however is, that should we by some wild happenstance still be around ten million years from now (average life span of a mammalian species is about two million years) I would expect us to have used our technology to better effect. More likely of course (and Ward addresses this possibility, but dismiss-es it) is that we will be replaced by the products of our technology long before then. Whether "they" will think it worthwhile to continue "living" is a very interesting question.

Clearly this is a popular book, almost a "coffee table" book, aimed at a popular readership, but that doesn't mean it's simplistic or dumbed down. True, Ward is biased toward a long-lived humanity which he thinks is likely the only intelligent creature in the cosmos (see *Rare Earth: Why Complex Life Is Uncommon in the Universe* (2000), which he wrote with Donald Brownlee), but Rockman's paintings really are first rate, and although the speculations are no more than that, they are interesting in themselves. Additionally there is a wealth of information in the text about evolution. Ward points out for example that it is not likely that we are going to undergo much Darwinian-type evolution in the future unless some humans become isolated. This can happen, he speculates, if an elite population isolates itself reproductively from the masses, or if we establish far-flung colonies in space. Another nice tidbit is Ward's observation that the average human I.Q. is not going to change much because whatever is measured on I.Q. tests is subject to the actions of numerous genes and any short term anomalies will be flooded by the mass of genetic humanity.

This book is a bit pricey because it is printed on expensive, glossy paper for the reproduction of the paintings. It's an attractive and entertaining book.

Wilson, Edward O. *The Future of Life* (2002) ****
A prescription offering hope

Famed biologist and godfather of sociobiology (and its current prodigy, evolutionary psychology), esteemed Harvard professor and one of the great scientists of our time, Edward O. Wilson outlines in this engaging but somewhat reserved book what is happening to the planet's biodiversity and what can be done about it.

The Prologue is a "letter" to Henry David Thoreau as Wilson seeks to establish a conservationist continuity between the author of *Walden* and ourselves. The open letter is somewhat self-conscious and artificial, but certainly appropriate for a work that celebrates nature and hopes to be a modest instrument in helping to preserve the natural world.

The first chapter is a survey of the life forms that live in "the biospheric membrane that covers Earth" (p. 21) with an emphasis on extreme climes including Antarctica's Lake Vostok (under two miles of ice) and the Mariana Trench (deepest part of the ocean at 35,750 feet below sea level). Chapter Two makes the assertion that the planet is currently going through a dangerous "bottleneck" characterized by disappearing habitats and extinction of species the likes of which have not been seen since the dinosaurs disappeared 65 million years ago. The culprit is of course us, represented by our short-sighted exploitation of non-renewable resources. Here Wilson begins his theme, to find a "universal environmental ethic" that will lead us "through the bottleneck into which our species has foolishly blundered." (p. 41)

In the next chapter, "Nature's Last Stand," Wilson delineates just how bad things really are as he surveys the rampant deforestation and other ecological obscenities currently taking place in the world. (Incidentally, those of you interested in a readable and painstakingly detailed account of what we are doing to mother earth, full of facts and figures, see Stuart L. Pimm's *The World According to Pimm: A Scientist Audits the Earth*, 2001.) Wilson continues with an estimate of how much the biomass is worth in dollars and cents ($33-trillion per year, which I think is similar to Pimm's figure). He makes the important point (which cannot in my opinion be repeated often enough) that the "cost" of doing business ought to include the damage or loss of "the free services of the natural economy" currently not figured into bottom line accounting. Thus the cost of extracting coal from the ground ought to include the value of the land torn up; the cost of wood from a tree ought to include the cost of watershed lost, etc. If the real costs of using the land, the rivers and the oceans, and the air were factored in—which some day they will be, whether we like it or not—some commodities would be seen as too expensive to harvest willy-nilly, and we might very well choose more environmentally agreeable alternatives.

In the final chapter Wilson gives "The Solution" which relies heavily upon the work of non-governmental environmental organizations that are attempting to use economic power to save the rain forests and other endangered "hotspots" throughout the world. Their technique includes outbidding the loggers for the rights to the forests, raising the standard of living of those who live in these endangered areas, and getting governments to see the value of their unspoiled lands.

Obviously Wilson is preaching to the choir here since myself and most others who will read this book will already be true believers in saving biodiversity. Perhaps the value of the book is in further educating us in the ways this might be done. Wilson is hopeful that we will wake up before it is too late. Indeed every minute counts because once the environment is gone it is gone forever to be replaced by God knows what. Wilson emphasizes not only the unknown value of all the plants, animals and microbes that are going extinct but the moral correctness of saving them. It is here that one notices a change in tone from the Edward O. Wilson of years ago. He is now so intent on saving what biodiversity is left that he is seeking to engage religion in the task!

This is Wilson somewhat mellowed at age seventy, seeking conciliation with former disputants for the greater good of planetary life. This is the entomologist as statesman.

The reason Wilson surprisingly points to the morality of saving wildlife as the key inducement is that we are robbing the world of our children and our grandchildren for our leisure and luxury today. It is a significant moral issue because we are putting what will be a terrible cost onto them, and they haven't a say in it at all!

I want to add that the danger inherent in the rampant devastation of the biosphere, whether through the direct destruction of ecologies or through pollution, is beyond our ability to foresee. The specter of a runaway greenhouse effect is just that, a phenomenon that may be upon us before we realize it, leaving us with no ability to stop it. Think of Venus and a surface temperature that melts lead. There is nothing in our present understanding of the biosphere that I know of that rules out that possibility. We are not only stupidly playing with fire, we are playing Russian roulette with ourselves and we are holding the gun to the head of our children. Wilson's book is an attempt to guide us away from such utter folly. I just hope that those people in the Bush administration and at the Wall Street Journal and the Economist and elsewhere who think that our resources will take care of themselves read this wise and penetrating critique and assume personal moral responsibility for their actions and utterances.

Index of Authors Reviewed

Index of Titles Reviewed

Why We Do It: Rethinking Sex and the Selfish Gene (2004), 101

Why We Get Sick: The New Science of Darwinian Medicine (1994), 73

www.ingramcontent.com/pod-product-compliance
Lightning Source LLC
Chambersburg PA
CBHW071415170526
45165CB00001B/287